CONCEPT BOOKS · 4

MATHEMATICS IN GENERAL

CONCEPT BOOKS

MATHEMATICS
IN GENERAL

A. E. F. DAVIS, M.A.
Senior Mathematics Master,
Bristol Cathedral School

HEINEMANN EDUCATIONAL BOOKS
LONDON

Heinemann Educational Books Ltd
LONDON   MELBOURNE   TORONTO
JOHANNESBURG   AUCKLAND   IBADAN
HONG KONG   NAIROBI   SINGAPORE

QA 39
.D29

SBN 435 46183 4

Published by Heinemann Educational Books Ltd
48 Charles Street, London W.1
Printed in Great Britain at the Pitman Press, Bath

To Giles-man, without whom this might have been
more quickly done.

Contents

Acknowledgements

It is difficult in a book based on casual talks to be sure acknowledgements are made. I have given references in the Notes to debts I am aware of, and apologize here to anyone unintentionally omitted.

I thank P. E. Parry and K. Jalie for reading, commenting on and checking the script and proofs, and am most grateful for the great care and trouble taken by the publisher and printer in setting the text.

31st December 1967   A.E.F.D.

Introduction

THE PRIME characteristic of an educated man is a proper curiosity. Without curiosity there can be no development of the mind. Education is 'leading out', a growth from present states and abilities to other and wider ones. This little book attempts to help the 'leading out' from basic school work in mathematics towards some awareness and understanding of what Mathematics is, in its many varied parts.

It is based on weekly talks given to sixth formers, both arts and science, over the past dozen years. Each chapter corresponds roughly to one week in a year's course. The amount of work needed to give benefit from a chapter varies greatly; there is always room for the extension of any topic which proves particularly interesting, and suggestions will be found in the text or notes.

I am not trying to teach anyone any mathematics, merely to help those who have been through part of a school course to see rather more of what Mathematics is about. The sixth former, student or general reader will have to *do* some mathematics and *find out* about mathematics in these pages and, I hope, will find it fun!

The reader must not expect his way to be smooth or to be fully detailed for him, only to find it sufficiently signposted. A signpost may be more intriguing than the road, or its end. A road passes through a district or a countryside; the interest may be *there*, unseen but for the sign!

Many mathematical terms have been italicised on first appearance in the text and their meaning, if not explained, should be clear from the context. A list appears at the end for reference. Notes are not referred to in the text but should be looked at after a first reading of a chapter.

A first reading should suffice to show whether the piece of mathematics seems worth pursuing: interests will vary as much as the topics. Symbols are the 'words' of the language of Mathematics; their extensive use in many chapters is intrinsic. I hope there is enough variety, however, to excite a useful interest and some understanding.

The progress of your enquiring mind is the measure of this book's worth!

1. Counting and Number Systems

MATHEMATICS MUST begin with counting. The first real Mathematics a child meets is counting. We shall see later that there is mathematics inherent in an ordered pattern that a child may make to fit coloured blocks into a box, and he may well do this before learning to count, but with counting comes the first contact with mathematical symbols. In a wider sense counting is the beginning of mathematics. The oldest pieces of mathematics extant are two Egyptian papyri of nearly 2000 BC—the Golenischev in Moscow and the Rhind in the British Museum. Both deal with the practical use of countable numbers. The former, and earlier, contains the solution of twenty-five problems, some involving implied knowledge of mensuration, whilst the papyrus in this country deals with eighty or ninety problems using fractions, equations and progressions as well as mensuration. If counting is obviously the start of practical mathematics, it is also the start of any attempt at the basic logical formulation of Mathematics. 'What is 1?' is a nice question. You may have fun discussing an answer! Unfortunately, the answers that mathematicians and philosophers have given, complete though they are, have not led to so comprehensive a build-up of mathematical structure as was once hoped.

Counting is the association of accepted word sounds and symbols with the common quality possessed by equi-numerate sets of objects; the common quality possessed by a man, a dog, a boat, a tree; by a pair of shoes, a couple of eggs, a brace of pheasants; and so on. It does not matter, intrinsically, what sounds or symbols are used, provided they are recognizable. You, yourself, cannot decide to say, 'An, deux, tre, four, funf, sex, sabah', because of the confusion your eccentricity would cause, but there is no *mathematical* reason why 1, 2, 3, 4, 5, 6, 7 should not be so linguistically mixed in their labelling.

In a more primitive form, however, counting may be purely perceptive; it can exist before any labelling by words or symbols.

1

Birds have been found to be able to 'count' to two, that is, they can distinguish between one object and two objects but then only see 'many' objects. Man's perception is similar but more advanced. Look out of the window at a group of people talking. How many are there? If there are two or three or four you answer at once; if five you may perceive this at once or almost unconsciously add two and three. If there are eight, say, you *cannot see* this but must consciously, albeit quickly, count them or add observed groups of perhaps three, two and three. Man, then, has an innate ability to count to four or five, after which all is *many*!

To represent three men by or six boats as

is little but pictorial shorthand and does not do much to further counting. To draw six boats as

does, however, further counting a little. The two sets of three boats help the eye and the understanding in a way the row of six boats does not. To consider III as a number quality possessed by the three men or $\frac{III}{III}$ as a number quality possessed by the six boats, is clearly a mark of a vastly higher degree of intelligence and shows an awareness of conscious counting. Let us fabricate a little history: suppose Man has reached the stage where he can represent the countability of objects by signs such as | or = or ∴.. He finds he cannot continue this. His Roman descendants will not write eight as IIIIIIII, nor the Chinese represent six as ☰ or the Maya use ∷ for seven. The eye and the sense is confused by the repetition as Alice was to be by the White Queen's question, 'What's one and one and one and one and one and one and one and one and one and one?'

New *patterns* of the symbol are needed, or *new symbols*. The use of single repeated strokes suggest the origin of some of the number symbols now in use but it is the need for further symbols to continue a system, that is really interesting. Find out for yourself, by reference to encyclopaedias or histories, the way various civilizations such as the Babylonian, the Mayan or the Chinese, dealt with the difficulty. Note the patterns of repeated symbols and the recurrent need for new hieroglyphs.

We may suppose a Roman IIII was followed by the sign of an open hand V for five. Numerals as far as eight or nine could then be represented by the reasonable maximum of four or five signs. A double hand spread $\times$ could lead to X for ten. These three symbols provide numbers up to about fifty. But there is then a stop! The counting must not stop! You know the subsequent signs that were used: L 50, C 100, D 500, M 1000, and many other forms. Find out the origin of these and the various ways that large numbers were written by the Romans. Clearly these large numbers defeat the purpose of a numeral representation. A number such as CIƆDCCCLXVII is not immediately understandable. It has to be built up in the mind piece by piece and thus is little more than a sort of abbreviation for the words which express the number.

Calculation in any such system is, of course, impossible without an abacus. The word itself means a sand covered table on which the finger can trace a pattern. Pebbles in parallel ruts of sand or beads on parallel wires make an easy way of representing a number, ready for addition or subtraction in trade dealings. An abacus of the type shown here is displaying MMMDLXII. Only five *calculi* are needed in each digit line, the upper groove showing nothing (down) or five (up), and the lower groove nought, one, two, three or four. Decide for yourself what operations would be necessary to add CDXXIX to the number on the abacus and then subtract MCCXLV from it to get the result MMDCCXLVI.

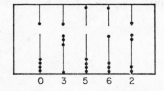

Various forms of the abacus have been used all over the world and are still extensively used. The Chinese are particularly adept at speedy calculation on an abacus small enough to be easily carried in the pocket. The virtue of an abacus is that the grooves or wires keep separate the digit lines, the fives, the tens, the hundreds and so on, in a way that the written number system did not often do.

We have glanced at Pictographic numeration and Hieroglyphic notation, the representation of number by pictures or by repeated symbols. Other ancient systems are alphabetical, notably the Hebrew and the Greek. If an alphabet is to be used to reach 999 there is a need for 27 letters, if there is no symbol for zero, 9 for the digits 1 to 9, 9 for the tens 10 to 90, and 9 for the hundreds 100 to 900. The Greek alphabet, however, contains only 24 letters, so three *old* letters were reintroduced into the alphabet for numerical purposes, *digamma* ς' for 6, *koppa* φ' for 90 and *sampi* λ' for 900. It is as if we, in England, reintroduced the Old English letters *ash* æ, *thorn* þ, and *yogh* ȝ, into our ABC and used them as numbers! The Greeks used dashes or bars to indicate that figures were being used and sub-dashes or further symbols for thousands and higher numbers. Thus we have, for example,

$$
\begin{array}{cccc}
\alpha' & \beta' & \gamma' & \delta'
\end{array}
\qquad \text{but} \qquad \overline{,\alpha\sigma\lambda\delta} \text{ for } 1234
$$
for $\quad$ 1 $\quad$ 2 $\quad$ 3 $\quad$ 4

Write out the Greek alphabet either in capitals or small letters—the small letters are a delight to write and beautiful to look at—and remember to put in the three extra letters. Discover for yourself the snags in this form of numbering. Clearly an abacus is needed for all calculation and clearly the counting is going to stop; at some stage the symbols are going to run out; new ones will have to be introduced.

No system we have looked at is complete: it is only practical with the aid of an abacus and can never be representative even of an Arithmetic of positive integers, for somewhere, the counting is going to have to end!

2. Positional Numeration: The Quinary Scale

A SMALL CHILD, and not only a small one, will use his fingers for counting. It is natural that man should always have used his fingers as an aid to counting and the manipulation of small numbers. Anthropologists know of the ways that primitive tribes use their hands to keep a tally, and you, yourself, may have seen the quick finger work which some Gypsies still use in their dealings.

Suppose a man starts counting with the little finger of the left hand. For convenience we shall use our own, 'One, two, three . . .' and '1, 2, 3 . . .' The man shows *one two three four* and can then continue in various ways. He may use the next finger as *five* and then go on to the next hand or repeat on the same hand, giving each ensuing finger a new name; or at some stage, at the end of one hand or two hands or of hands and feet, he will call a stop and begin again with a new labelling. If after *one two three four* he shows the full single hand and says, '*A span*', he can continue, *span and one*, *span and two* and so on. If at a display of both hands he announces, '*A span*', he goes on as before *span and one*, *span and two* . . . but, of course the *numbers* so labelled are not the same as before. We can suppose a *span* to be at any convenient point where the 'digits' are, for the moment, complete—at the five 'digits' of one hand, at the ten of both hands, at the twenty of hands and feet; or after the four fingers of one hand or the eight fingers of both; or theoretically, if not historically, after any number of 'digits' decided upon as a *span*.

A pattern of counting has now evolved. How far it would go in practice would depend upon factors other than the mathematical, but the pattern is established. Two lines of development are possible. A *span* represented as V leads to VI etc. to X for a double *span* and to all the awkward incompleteness of the Roman system or many another using its own symbols. Alternatively, *one two three* . . . are at first labelling the fingers; why should they not subsequently label the *spans*? The sequence will go:

5

one two three . . .	*one* span
one span and one and two . . .	*two* spans
two spans and one and two . . .	*three* spans
. . .	*a span* of spans!

This counting has no end; it has no stop!

This movement of counting on the fingers can be compared with movement of beads on an abacus. Consider this carefully for yourself.

The abacus suggests a source for a numeration which is not incomplete. A symbol is needed for a line of the abacus which shows *no* beads. Only the ancient peoples of the Mayan civilisation and, more important to our intellectual and economic history, the Hindu-Arabian, developed the brilliant concept of a symbol for *nothing*. The Arabic *cifa* is as fundamental to arithmetical progress as the invention of the wheel is to the history of transportation. It was not until after the fall of Constantinople in 1453 and the western flow of Greek and Arab learning, and with the vast increase in trading, in navigation and in exploration, that gradually the awkward Roman numbers were displaced by the Arabic. There is a Latin manuscript of AD 967 in the Escorial Library containing some of these numerals and there are occasional examples of their use, especially in Italy and Spain, through earlier Moorish influence, before their more general use in England and Europe from the sixteenth century onwards.

The 0 makes possible *positional numeration*. For the 0, admittedly meaning *nothing*, meaning an empty line on the abacus, can denote what I have called a *span*. The counting is now:

1	2	3	. . .	10
11	12	13	. . .	20
21	22	23	. . .	30
		. . .		100 and so forth for ever.

Our system of counting is Arabic because of this all powerful *cifa*. It is not primarily Arabic because of any resemblances between the shape of the Arabic symbols and our own: the resemblances are few and without much significance. (You may, however, find it of interest to learn about the history of our numerals.)

We go back to the man counting on his left hand, and suppose he is using only the one hand. We shall call his counting Quinary— for obvious reasons—and see how it works.

Head a piece of paper, Quinary. Using the fingers of one hand for counting, write down with the other, to-day's date and your age. For me the answers are:

To-day is the 21st day of the 11th month.
My age is 131 years and 20 months.

Practise counting on your fingers *one two three four ten eleven twelve thirteen fourteen twenty twenty-one twenty-two twenty-three twenty-four thirty . . . forty-four hundred hundred and one . . .* until all this becomes easy and clear.

You now have learnt to count in Quinary. Years ago you learnt to count in Denary—the usual *scale of notation*—with both hands denoting the *span*. Now learn to add and to subtract. When you 'carry' or 'borrow' you are still transferring a *ten* but the value is that of the digits of but one hand. You should find this easy; if you don't, use your fingers—or mentally cheat by thinking of carrying or borrowing a 5! Check that the examples below are right and make up and check some examples of your own.

$$
\begin{array}{r}
3043 \\
231 \\
10 \\
44013 \\
32 \\
\hline
102434
\end{array}
\qquad
\begin{array}{r}
1343 \\
-420 \\
\hline
423 \\
\text{check} \quad 420 \\
\hline
1343
\end{array}
\qquad
\begin{array}{r}
430213 \\
-43224 \\
\hline
331434 \\
\text{check} \quad 43224 \\
\hline
430213
\end{array}
$$

You next must learn your Tables! Otherwise you will be forced to do all multiplication by continued addition and all division as continued subtraction. The Tables of multiplication are easy, as clearly, the highest multiple that is needed is four times four, that is 4 . 4. Here is the Table—omitting the obvious:

$$2 . 3 = 11 \qquad 3 . 3 = 14$$
$$2 . 4 = 13 \qquad 3 . 4 = 22 \qquad 4 . 4 = 31$$

Check these calculations and do some examples of your own making.

$$12 . 3 = 41 \qquad \frac{24}{2} = 12 \qquad 3\overline{)4122} \atop 1204$$

$$31 . 4 = 224$$

$$333 . 2 = 1221 \qquad \frac{34}{2} = 14\tfrac{1}{2} \qquad 4)\overline{1234}$$
$$2 . 2 . 2 . 2 . 2 = 112 \qquad\qquad\qquad 143 \text{ rem. } 2$$

(Query: Are there odd and even numbers in Quinary?)

$$
\begin{array}{r}
1324 \\
203 \\
\hline
320300 \\
10032 \\
\hline
330332
\end{array}
\qquad
\begin{array}{r}
1\cdot 03 \\
12)\overline{13\cdot 000} \\
12 \\
\hline
100 \\
41 \quad \text{etc.} \\
\hline
\end{array}
$$

You now have the basic arithmetic of the Quinary scale. What advantages and disadvantages has it over our Denary scale?

Finally, can you complete to-day's date and state the year?

3. More Non-Denary Arithmetic

AN ATTEMPT was made in the last chapter to show how a decimal system of counting could have arisen. By a *decimal* system is meant a positional system depending upon a 10. Thus the Quinary system— a single hand—is decimal and so is our usual Denary scale of notation—both hands. Notice the distinction between the meanings of decimal and Denary. The essence of Arabic numeration is that it is decimal; that with a *limited* set of defined symbols and a 0, there is *unlimited* counting: in fact, of course, it is Denary.

You have had some practice in counting in fives. Historically counting has usually been in fives or multiples of fives, although, as we have seen, it has not usually been decimal. An Arab people, the Nabataeans, apparently counted in fours, and our word *eight* comes from Sanskrit *two fours*: British units show some dependence upon twelves. The Babylonians and Egyptians counted in sixties, a sixty, you notice, containing both five and twelve, perhaps related to a yearly cycle of roughly 360 days, and giving us our time and angle measurements. Mayans, Aztecs and Eskimos counted in twenties; the French language retains traces of counting in twenties and the Welsh language of counting in fives. The Chinese, the Hebrews, Greeks, Romans, Arabs, all, in their several ways, used the ten we know.

Our purpose, now, is to look more closely into the structure of decimal counting by considering further *tens*. But first, what about the year, in Quinary? The meaning of 1966, for instance, is

$$1 \cdot 10^3 + 9 \cdot 10^2 + 6 \cdot 10 + 6$$

where the 10 is Denary. We need to re-express this number, or a similar one, in fives. Dividing by 5 we have

$$5)\overline{1966}$$
$$\overline{393 \text{ rem. } 1}$$

This tells us that there are 393 Quinary *tens* in 1966 and 1 *unit* over. Dividing again:

$$5)\overline{393}$$
$$\overline{78 \text{ rem. } 3}$$

that is, 78 Quinary *hundreds* and 3 *tens* over. Repeating this and completing it, the successive divisions give:

	Denary		Quinary
5	1966		
5	393	remainder	1 unit
5	78	remainder	3 tens
5	15	remainder	3 hundreds
5	3	remainder	0 thousands (note this!)
	0	remainder	3 ten thousands
	1966	←⊢→	30,331

1966 in Denary has been expressed as

$$3 . 5^4 + 0 . 5^3 + 3 . 5^2 + 3 . 5 + 1$$

that is, 30,331 in Quinary.

Now practise the method of conversion, turning Denary numbers into other Scales; compare your results with actual counting. For instance:

There are 11 matches in my match box. How many is this in the Scale of three?

(i) By counting: (ii) By division:

Denary	Ternary
1	1
2	2
3	10
4	11
5	12
6	20
7	21
8	22
9	100
10	101
11 →	102

	Denary	Ternary
3	11	
3	3	2
3	1	0
	0	1
	11 →	102

Thus, in the Ternary Scale there are 102 matches in the match box.

If you by now understand positional numeration, the reverse process to this should be obvious. Practise it. On the next page is an example from Scale eight.

$$
\begin{array}{lll}
\text{Octal} & \text{Denary} \\
1\ 3\ 5\ 7 \longrightarrow & 7 \cdot 1 = & 7 \\
\qquad\quad \longrightarrow & 5 \cdot 8 = & 40 \\
\qquad\quad \longrightarrow & 3 \cdot 8^2 = 3 \cdot 64 = & 192 \\
\qquad\quad \longrightarrow & 1 \cdot 8^3 = 1 \cdot 512 = & 512 \\
\hline
& & 751
\end{array}
$$

A shorter way is to reverse the division method:

$$
\begin{array}{lll}
\text{Octal} & 1\ 3\ 5\ 7 \\
\times 8 & 11 \\
\times 8 & 93 \\
\times 8 & 751 & \text{Denary,}
\end{array}
$$

where the arrowed digit is added after each multiplication.

$$* \qquad * \qquad *$$

After learning to count in Quinary you went on to the basic arithmetic in that Scale, remembering that the 'carried' or 'borrowed' *ten* had its *radix* value, i.e. its Scale value. You next needed the multiplication Tables and were then able to go on to multiplication and division. Try out all these things in any Scale of your own choosing. Here is some material in the Scale of six.

(i) Counting.

$$
\begin{array}{cccccc}
1 & 2 & 3 & 4 & 5 & 10 \\
11 & 12 & 13 & 14 & 15 & 20 \\
21 & 22 & 23 & \ldots & & \\
& & \ldots & 54 & 55 & 100 \\
101 & & \ldots & & &
\end{array}
$$

(ii) Addition and subtraction.

$$
\begin{array}{rr}
240 & 34501 \\
1235 & -24343 \\
10 & \rule{1.5cm}{0.4pt} \\
4413 & 10114 \\
55 & \text{check}\ \ 24343 \\
\hline
10441 & 34501
\end{array}
$$

(iii) Tables.

$$2.3 = 10 \qquad 3.3 = 13$$
$$2.4 = 12 \qquad 3.4 = 20 \qquad 4.4 = 24$$
$$2.5 = 14 \qquad 3.5 = 23 \qquad 4.5 = 32 \qquad 5.5 = 41$$

(iv) Multiplication and division.

$$34.5 = 302 \qquad\qquad 5)\overline{12345}$$
$$3^3 = 43 \qquad\qquad\qquad 1421$$

*   *   *

We have considered so far Scales with a radix lower than our own. What of the Duodenary Scale? In this, most conveniently, there would be 16 pence in 1s. 6d. and 34 inches in 3 ft. 4 in. There would be 10 months in a year. But can *you* count to 20 with this twelve base? Try it!

.   .   .   .

1	2	3	4	5	6	7	8	9			10
11	12	13	14	15	16	17	18	19			20

Clearly there must be additional symbols and words for the unlabelled numbers. Use what you will! I shall use T tee, E ee, 1T teeteen and 1E eeteen.

Do some Arithmetic in this Scale: I don't suppose you will fancy attempting the Tables—from 2 . 2 to E . E demands a lot of effort.

Note that the decimal fractions are interesting, as division by 2, 3, 4 and 6 will always give simple terminating decimals.

$$\tfrac{1}{2} = \cdot 6 \qquad \tfrac{1}{3} = \cdot 4 \qquad \tfrac{1}{4} = \cdot 3 \qquad \tfrac{1}{6} = \cdot 2$$

There is much talk about the change to decimal currency and the Metric system. It is inconceivable that the world would change, instead, to a Duodenary numeration, but would this be, were it possible, a good thing?

4. The Binary Scale

POSITIONAL notation for numbers requires a finite set of symbols which include the zero symbol. There must be at least one symbol other than the zero, so that one *span* can exist—so that there is a radix. The 'simplest' counting, therefore, will use only 1 and 0. This could be said to derive from counting on the thumbs alone: you may try this! Clearly it is not practicable nor historically reasonable. The radix *ten* is reached too frequently to be useful on the hands.

The Binary scale is, however, of importance just because it is, nominally, the easiest system; because it employs only two symbols and has only two possible operations. The addition table of digits shows that two *like* symbols produce 0 and two *unlike* produce 1.

$$
\begin{array}{c|cc}
+ & 0 & 1 \\
\hline
0 & 0 & 1 \\
1 & 1 & 0
\end{array}
$$

We shall see that the $+$ operator of the table suffices for the arithmetical operators.

A very little practice should make Binary counting straightforward:

1	10				
11	100	101	110		
111	1000	1001	1010	1100	1101
1111	10000	and so on . . .			

Convert small numbers to Binary by counting and try changing large numbers by the division method, taking out twos, that you already know. Also choose some number, made up at random, such as 11011001101 and find out its value in Denary, 1741, either by recalling the positional meaning of the unity digits, $2^0\ 2^1\ 2^2\ 2^3$ . . . that is 1 2 4 8 . . . reading from right to left, or by reversing the division method as shown in the previous chapter.

13

Do some addition—you will find it easy but laborious—and some subtraction. Check first the examples below and then make up your own. You will find it helpful to write Binary digits in blocks of three and to read them as digits.

```
   1 101 011           10          101 101
 +   110 100         −  1        − 011 011
 ─────────────      ──────       ─────────
  10 011 111            1          010 010
```

```
       101 011              101 011 100
       011 110            − 011 001 111
       111 101            ─────────────
       011 110              010 001 101
     ─────────
     010 100 100
```

The only product in the multiplication table is obviously 1 . 1 so there is no multiplication or division in Binary. 'Multiplication' by 1 is *writing down*, 'multiplication' by 0 is *not writing down*. Multiplication is addition and division is subtraction.
Examples: (i)

Multiply 110 010 by 1 001
Product is   110 010 000 'thousand times'
                              +   110 010 'one times'
                                        ─────────────
                                        111 000 010

(ii) Divide 011 110 by 101

```
                110
       101)011 110
           10 1
           ─────
            1 01
```

We now have a method for doing Denary multiplication entirely by very easy addition. Let us evaluate 83 × 19.

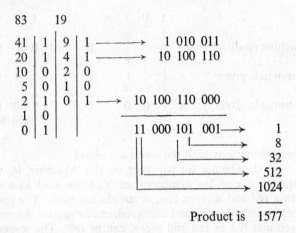

83		19		
41	1	9	1	———→ 1 010 011
20	1	4	1	———→ 10 100 110
10	0	2	0	
5	0	1	0	
2	1	0	1	———→ 10 100 110 000
1	0			
0	1			

11 000 101 001 ——→ 1
                → 8
                → 32
                → 512
                → 1024

Product is 1577

Try a multiplication of your own and discuss whether this is a worth while method is practice.

If possible, get a line of people to stand side by side in front of you, left arms down and outstretched right hands resting on the next's shoulder. Each can represent a 0; the raised *right arm* will represent a 1. We have here a human picture of a Binary number.

Start with each person in the zero position. There are just *two* rules for counting on the human calculator. An instruction is received by a hand falling upon the shoulder. Then

   (i) if your arm is *down*, put it *up*
   (ii) if it is *up*, put it *down*.

Remember you do nothing unless commanded by the fall of a hand onto your shoulder. The first digit can be activated by words from the 'Machine Operator'. At each *one two three* . . . it responds as to a touch of a hand. The Binary forms of the *one two three* . . . should then appear in the show of arms.

Let 13 be multiplied on the machine. We have 101 × 1101.

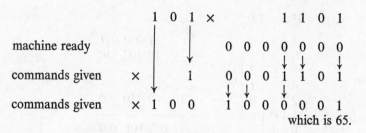

which is 65.

Try other calculations until the process is clear!

The method, whether on paper or on the 'Machine' is, with a little experience, easy but cumbersome. You can work in a similar way with a red and a green flag, or an electric bulb. The response to a green and a green or a red and a red, can be green: the response to a green and red or red and green, can be red. The response to bulb off and off or on and on, can be off: that to off and on or on and off, be on. In each case we have the same behaviour pattern:

and	green	red		+	0	1		and	off	on
green	green	red		0	0	1		off	off	on
red	red	green		1	1	0		on	on	off

It is easy, slow and laborious. But what of electronic impulses acting according to the 'off-on' table half-a-million times a second? The cleverness of the electronic computer lies in the ingenuity of its builder and not in the basic way it works. The addition of two Binary digits is an extremely simple operation: it is the brilliant construction of instruments capable of repeating such an operation at unimaginable speeds, that makes the computer the amazing, though essentially stupid, thing that it is. This is, of course, but a small part of the story: much of the value of a computer depends upon its ability to store information.

Multiplication is addition and division is subtraction, but we have not yet tied the operation of subtraction to the + operator of the table. This is done by considering the *complement* of the number to be subtracted. The *complement* of a digit in any Scale, is the number which brings that digit up to the highest digit in the Scale. Here we have only to interchange the 0's and 1's. Look back at the third

subtraction example. The complement of 011 001 111 is 100 110 000
and

$$101\ 011\ 100$$
$$\text{plus}\quad 100\ 110\ 000$$
$$\text{is}\quad 1\ 010\ 001\ 100$$

To get the difference we want, from this sum, we *add one* to the
right hand digit and *then* remove the left hand digit. This gives
010 001 101 the correct answer. It seems a strange method! See
if you can justify it. The method may be used in any Scale of
notation. To get the complement of a number, subtract it from the
highest numeral in the system: thus the complement of 3 in Quinary
is 1 and the complement of Denary 2 is 7. Try ordinary subtraction
in this way: make the number of digits the same in both numbers
chosen, by the addition of noughts where necessary.

*   *   *

You have probably heard of Russian Peasant multiplication.
Two numbers to be multiplied are put side by side and the smaller
successively halved without remainder, whilst the other is doubled.
Thus, taking the example earlier in the chapter,

$$83 \times 19$$

166	9
~~332~~	~~4~~
~~664~~	~~2~~
1328	1

All *evens* of the smaller column
are erased and the answer got by
addition.

1577

Try out the method, if you don't already know it. Compare
the working above with the Binary working and discover for yourself
the secret of Russian multiplication!

5. Algebras: (a) Lewis Carroll's Implication Problems

CHARLES LUTWIDGE DODGSON was nominated, at the age of twenty, Student of Christ Church, Oxford in 1852. He was to be a lecturer in Mathematics for nearly thirty years.

Apart from his Lewis Carroll fantasies, he published, among his mathematical works, a book on 'Symbolic Logic'. Many of his examples in this book make a light-hearted introduction to non-numerical Algebra.

Let letters $A, B, C \ldots$ stand for single statements such as

$$R = \text{It is raining.} \qquad C = \text{I put on my coat.}$$

These are simple sentences, meaningful and declarative, which can be classified, without qualification, as either True or False.

Consider the sentence, 'If it is raining, I put on my coat'. This can be separated into two *connected* statements:

$$\textit{If} \text{ it is raining, } \textit{then} \text{ I put on my coat,}$$
$$\text{or } \textit{if} \qquad R \qquad \textit{then} \qquad C$$

or symbolically $R \Rightarrow C$ where $\Rightarrow$ is read 'implies', but is understood to have here the specific meaning of, if . . . then . . .

Dodgson's problems can be reduced to a series of sentences of this type. The negative of a statement A shall be written A' and read *not A*: for instance, $R' =$ It is not raining.

One problem, slightly adapted, reads:
(1) Every one who is sane can do Logic.
(2) No lunatics are fit to serve on a jury.
(3) None of *you* can do Logic.
What conclusion may be drawn?

Choose symbols which are suggested by the sentences:
(1) The sane can do Logic. $S \Rightarrow L$
(2) If not sane, then no jury service. $S' \Rightarrow J'$
(3) *You* cannot do Logic. $U \Rightarrow L'$

For a conclusion to be reached, there must be a chain of implication, such as $A \Rightarrow B \Rightarrow C \Rightarrow \ldots \Rightarrow K$ so that, finally, $A \Rightarrow K$. There is no such obvious sequence in the above sentences.

Let us, for the moment suppose we have only the earlier R and C. We are given $R \Rightarrow C$. Accepting this as *true*, are other possible relations between R and C necessarily true? Think about each of these:

$R \Rightarrow C$	If it is raining I put on my coat.
$C \Rightarrow R$	If I put on my coat it is raining.
$R' \Rightarrow C'$	If it is not raining I do not put my coat on.
$C' \Rightarrow R'$	If I do not put my coat on it is not raining.

Assuming a world of statements A and B, *only*, what follows from the one implication $A \Rightarrow B$? We have:

the Sentence $A \Rightarrow B$	its Converse $B \Rightarrow A$
its Inverse $A' \Rightarrow B'$	and its Negative $B' \Rightarrow A'$

Choose any A and B you like and convince yourself that if the Sentence is *true* then only its Negative is consequentially true.

We now consider $A \Rightarrow B$ and $B' \Rightarrow A'$ as equivalent, that is, each necessarily implies the other, and interchange them at will.

Go back to the problem. The implications are now:

$S \Rightarrow L$ or $L' \Rightarrow S'$	Consider the three $L' \Rightarrow S'$
$S' \Rightarrow J'$ or $J \Rightarrow S$	$S' \Rightarrow J'$
$U \Rightarrow L'$ or $L \Rightarrow U'$	$U \Rightarrow L'$

These last three at once lead to $U \Rightarrow L' \Rightarrow S' \Rightarrow J'$ or $U \Rightarrow J'$, namely, *You* are not fit to serve on a jury! Alternatively, $J \Rightarrow S \Rightarrow L \Rightarrow U'$ gives $J \Rightarrow U'$, A jury will not have *you* upon it!

Here is a further example:

(1) Promise-breakers are untrustworthy.	$P' \Rightarrow T'$
(2) Wine-drinkers are very communicative.	$D \Rightarrow C$
(3) A man who keeps his promises is honest.	$P \Rightarrow H$
(4) No teetotalers are pawnbrokers.	$D' \Rightarrow B'$
(5) One can always trust a very communicative person.	$C \Rightarrow T$

The symbols should explain themselves. P keeps promises, T trustworthy, D drinker, H honest and B broker. There are two *loose* symbols H and B. We start either with H or with B and attempt a chain of implication, remembering that any implication may be replaced by its negative.

Either $H' \Rightarrow P' \Rightarrow T' \Rightarrow C' \Rightarrow D' \Rightarrow B'$ that is $H' \Rightarrow B'$,

A dishonest man is not a pawnbroker;

or $B \Rightarrow D \Rightarrow C \Rightarrow T \Rightarrow P \Rightarrow H$ which is $B \Rightarrow H$,

A pawnbroker is an honest man.

Lewis Carroll's own answer was: No pawnbroker is dishonest.
Try the following:

(i) (1) Babies are illogical.
   (2) Nobody is despised who can manage a crocodile.
   (3) Illogical persons are despised.
(ii) (1) No one takes the *Times* unless he is well educated.
   (2) No hedge-hogs can read.
   (3) Those who cannot read are not well educated.
(iii) (1) Nobody, who really appreciates Beethoven, fails to keep
         silent while the Moonlight Sonata is being played.
   (2) Guinea pigs are hopelessly ignorant of music.
   (3) No one, who is hopelessly ignorant of music, ever keeps
         silent while the Moonlight Sonata is being played.
(iv) (1) No boys under 12 are admitted to this school as boarders.
   (2) All the industrious boys have red hair.
   (3) None of the day boys learn Greek.
   (4) None but those under 12 are idle.
(v) (1) No kitten that loves fish is unteachable.
   (2) No kitten without a tail will play with a gorilla.
   (3) Kittens with whiskers always love fish.
   (4) No teachable kitten has green eyes.
   (5) No kittens have tails unless they have whiskers.
(vi) (1) I never put a cheque, received by me, on that file, unless
         I am anxious about it.
   (2) All the cheques received by me, that are not marked with
         a cross, are payable to bearer.
   (3) None of them are ever brought back to me, unless they
         have been dishonoured at the Bank.
   (4) All of them that are marked with a cross are over £100.
   (5) All of them not on that file, are marked 'not negotiable'.
   (6) No cheque of yours, received by me, has ever been dis-
         honoured.

(7) I am never anxious about a cheque, received by me, unless it should happen to be brought back to me.
(8) None of the cheques received by me, that are marked 'not negotiable', are for over £100.

6. Algebras:
(b) What is Algebra?

THE ALGEBRA of Implication is a simple Algebra. Algebra can be compared to a Game. A game requires 'men'—real men in Rugby, counters in Monopoly, cards in Bridge—with which to play; operations they can perform—a touch down in Rugby, buying a house in Monopoly, trumping an ace in Bridge—and, of course, Rules. With these essentials, Men and Operations and Rules, a game can be made; whether a particular game constructed on these lines is worth playing is quite a different matter! A game does not cease to be a game because it is a poor one or because one does not like it.

Solving Lewis Carroll's Logic problems was like a game. The 'men' played with were the Statements—'no hedgehog can read', 'all industrious boys have red hair'—and the operations depended upon ⇒. The rules were to attempt to find a chain sequence of Implication and that the negative consequence of $A \Rightarrow B$ is $B' \Rightarrow A'$. Let us put this more formally. The Algebra of Implication has:

Symbols A, B, C . . . for single statements
Operators   ⇒   for implication, if . . . then . . .
Rules   1. $(A \Rightarrow B) \wedge (B \Rightarrow C) \Rightarrow (A \Rightarrow C)$
       2. $(A \Rightarrow B) \Leftrightarrow (B' \Rightarrow A')$

This looks terrifying! Read these rules like this. 1. If statement A implies statement B and B implies C then A implies C. 2. The sentence 'A implies B' implies, and is implied by the sentence 'not-B implies not-A'.

The introduction to this in the last chapter was easier but the Symbols, Operators, Rules formulation is the mathematical expression of the Algebra. An Algebra is an analysis using stated Symbols behaving with stated Operators according to stated Rules. An Algebra may be fun or a bore, useful or useless, a mathematician's delight or a practical tool (or both!) but it is an Algebra regardless of these qualifications.

'School Algebra' is Number Algebra. Let us pretend at the age of eleven you go into class and the teacher begins, 'Today we start Algebra. We have:

> Symbols $a, b, c, \ldots$ for numbers.
> Operators   Addition and multiplication with their inverses subtraction and division.
> Rules   1.   Addition and multiplication are Commutative
> $a + b = b + a$ and $a \cdot b = b \cdot a$
>   2.   Addition and multiplication are Associative
> $a + (b + c) = (a + b) + c$ and $a \cdot (b \cdot c)$
> $= (a \cdot b) \cdot c$
>   3.   Multiplication is Distributive over Addition
> $a \cdot (b + c) = a \cdot b + a \cdot c$ but
> Addition is not Distributive over Multiplication
> $a + b \cdot c \neq (a + b) \cdot (a + c)$

You now know Number Algebra."

This teacher's approach is ridiculous, of course! No child will learn much from so axiomatic a beginning. I cannot imagine many people learning Rugby, Monopoly or Bridge by reading the Rule books! One learns a game by playing it: one learns an Algebra by practising it. Go back to the Rules given above and convince yourself nonetheless of their truth by using whole numbers. Try to remember when you intuitively picked up these facts although you never had them stated or named.

If then we are going to look at some other Algebras we may glance at the mathematical formulation under the headings of Symbols, Operators, Rules and see something of the behaviour of that Algebra but we won't be able to learn to employ or understand it: it needs a lot of practice, that is, playing the game of that Algebra.

Consider the following formulation:

Symbols X means Mr. Exe is telling the truth.
Y means Mrs. Why is telling the truth.
Z means Miss Zed is telling the truth.

Operators   ° means 'and'   * means 'or'.
Rules   Discover for yourself whether the Algebraic Laws of Commutation, Association and Distribution hold. Look at these propositions

$$(c)\ X°Y \overset{?}{=} Y°X \quad X*Y \overset{?}{=} Y*X$$
$$(a)\ X°(Y°Z) \overset{?}{=} (X°Y)°Z \quad X*(Y*Z) \overset{?}{=} (X*Y)*Z$$
$$(d_1)\ X°(Y*Z) \overset{?}{=} (X°Y)*(X°Z)$$
$$(d_2)\ X*(Y°Z) \overset{?}{=} (X*Y)°(X*Z)$$

Put these into words and decide for yourself if each is true.

If it is, replace $\overset{?}{=}$ by $=$, and if not by $\neq$. (c) expresses the Commutative Law, (a) the Associative Law, (d_1) the Distributive Law for the operation ° over the operation * and (d_2) that for * over °.

You now have a New Algebra! Whether it is a game worth playing is another matter! Compare X, Y, Z with whole numbers x, y, z and °, * with addition and multiplication and so discover in what ways Number Algebra and this New Algebra are similar and dissimilar.

You may consider the New Algebra trivial but you will have found it (I hope) to have a neat structure. Set Algebra has a similar structure and we will look at this more fully.

A Set or Class is any specified collection brought together— {rusty bicycles in Hong Kong} or {even numbers} or {Tom, Dick, Harry} or $\{A, \Omega\}$ or {rectangles} or {the dramatic Unities} or $\{1, \frac{1}{2}, \frac{1}{4}, \frac{1}{8}, \ldots\}$. A capital letter will represent a Set and we will diagrammatically put the elements of a Set in a hoop thus:

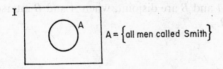

where the outer shape I is the *Universal set* of things under discussion, perhaps Englishmen in this case or the men listed in the London telephone directory. Choose any Universal set you wish such as Cars or Books in the Library or Women and draw diagrams of this

type with *two* Sets shown within I. Consider carefully in what hoop
or in what *part* of a hoop each element or member of a Class must be.
Here are two examples:

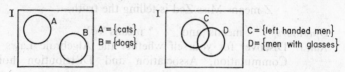

         The Sets are *Disjoint*   The Sets *Intersect*

Be sure you draw some diagrams where one Set is a *Subset* of the
other, such that *all* the members of one Class are also members of
the other. You may also try your ingenuity in determining the Venn
diagram (as these pictures are called) for more than two Sets within
I. Suppose the Universal Set consists of plane quadrilaterals and
let $A =$ {trapezia} $B =$ {rhombi} $C =$ {rectangles} $D =$ {squares}
$E =$ {cyclic figures}; or consider your friends and the Sets of those
who speak some French, German, Spanish, Russian.

We will see later that these sketches themselves can be used to
solve certain problems concerning the number of elements in
particular Classes but the present purpose will be to find out something of an Algebra using these Set symbols.

We have so far Symbols but neither Operators or Rules. Three
operators need to be defined. The *Union* of sets A and B is written
$A \cup B$ and read 'A union B'. It is {all elements which are members
of A *or* B *or* of both}. The *Intersection* of sets A and B is written
$A \cap B$ and read 'A intersection B'. It is {all elements which are
members of both A *and* B}. The *Complement* of Set A is written A'
and read 'not-A'. It is {all elements *not* members of A}.

Try to draw Venn diagrams of $A \cup B$ and $A \cap B$ in the three
cases, when A and B are disjoint, when A and B intersect, when A is
a subset of B.

7. Algebras:
(c) Set Algebra; Boole

YOU HAVE met three operators in Set Algebra. Remind yourself

Union $A \cup B$ Intersection $A \cap B$ A', Complement of A

of these operations of Union, Intersection and Complementation by drawing Venn diagrams of the following: $(A \cup B)'$, $(A \cap B)'$, $A' \cup B$, $A \cap B'$, $A' \cup B'$, $A' \cap B'$. Two are shown here. Choose

A is a sub-set of B. $A \subset B$ $A \cap B$ is not empty. $A \cap B \neq \phi$
      $(A \cup B)' = B'$ $A \cap B' = A - B$

particular sets for your diagrams and decide which elements belong to the shaded set. Wherever possible find an alternative symbolic way of referring to the set. In the first diagram above Set A has been chosen so that all its members are also members of B: A is *included* in B: thus $A \cup B = B$ and so $(A \cup B)' = B'$. Already we are *doing* a little Algebra. Without reference to a diagram, the last sentence could have been written as

$$(A \subset B) \Rightarrow A \cup B = B \Rightarrow (A \cup B)' = B'$$

In the other diagram A and B have been chosen so that there are some elements common to both sets. $A - B$ seems a 'common-sense' expression for the picture: again, a little Algebra! We could *define* the operation of Set subtraction by this, $A - B = A \cap B'$.

25

If two sets have no common members there is no intersection in the diagram but for algebraic completeness we say the set $A \cap B$ is the *empty* set $\emptyset$ which has *no members*. Do not say the empty set contains nothing; the set $\{0\}$ contains one element, nothing: it is not empty!

Convince yourself of the correctness of these expressions:

$$A \cap A' = \emptyset \qquad A \cup \emptyset = A \qquad A \cap \emptyset = \emptyset$$
$$A \cup A' = I \qquad A \cap I = A \qquad A \cup I = I$$

Our Set Algebra now has its Symbols and its Operators and it will be possible to see next how it behaves in its Rules—we have met quite a few things already.

It is evident without demonstration that both Union and Intersection are Commutative. Are they Associative? Convince yourself that they are, that if A, B and C are connected by two operations of Union, or by two operations of Intersection, the resulting Set is the same whatever the way of working. Brackets are therefore not needed when Union is repeated or when Intersection is repeated.

$$A \cup B \cup C \qquad\qquad\qquad A \cap B \cap C$$

Is there a defined meaning if the operations are mixed? Does $A \cup B \cap C$ have a unique meaning? Does $1 + 2 \cdot 3$ have a unique meaning?

You will find the meaning of $A \cup B \cap C$ to depend upon the order of working. As the operations are Commutative $(A \cup B) \cap C = C \cap (A \cup B)$ and $A \cup (B \cap C) = (B \cap C) \cup A$ but the two pairs are *not* the same.

Let us for the moment leave Set Algebra and suppose you are given a pot of paint and told,

'*You* paint the table and the chair!'

or $\qquad U_p(T \wedge C)$ where the symbols explain themselves.

The meaning of the instruction is

You paint the table and *you* paint the chair,

or $$(U_pT) \wedge (U_pC)$$

This suggests:

$$U_p(T \wedge C) = (U_pT) \wedge (U_pC)$$

Now try this on $A \cup (B \cap C)$. Draw Venn diagrams to demonstrate

$$A \cup (B \cap C) = (A \cup B) \cap (A \cup C)$$

and then try the *dual* form for $A \cap (B \cup C)$.

You have found that each operation of Union and Intersection is Distributive over the other. Make up examples and practise the use of the Law with such expressions as $P \cap (Q \cup R)$, $(X \cap Y) \cup Z$, $(L \cup M) \cap (L \cup N)$, or $(A \cap B) \cup (B \cap C)$.

Simplify the following in any way possible:

   (i) $(A \cap A) \cup A$
  (ii) $A \cap (A \cup B)$
 (iii) $A \cup (B \cup C)$
 (iv) $(A \cap B) \cup \emptyset$
  (v) $I \cap (A \cup A)$
 (vi) $(A \cup \emptyset) \cap (A \cup I)$
(vii) $(A \cap A) \cup (A \cap B) \cup (A \cap C)$
(viii) $(A \cap B) \cap C$
 (ix) $(A \cap B) \cup A$

We have, by now, a fair amount of algebraic material. Some of it has arisen, incidentally, from the concept of a Set or from Venn diagrams; some of it expressed in behavioural Rules. The operators $\cup$ and $\cap$ seem *algebraically* very similar and independent of each other. We are going to see, however, that they are interdependent under Complementation.

Consider two sets of men: those who have cars and those who have children. How can we represent the set of those with neither? The word *neither* suggests the Complement of Union, $(X \cup Y)'$ say, and also, *not cars* and *not children*, $X' \cap Y'$. This leads to de Morgan's Law $(X \cup Y)' = X' \cap Y'$ and its *dual* $(X \cap Y)' = X' \cup Y'$. Draw diagrams to make this clear.

To test the completeness of our labelling—to see that every set within a system has a label—try naming the 16 sets in the diagram. The small letters can be used as a shorthand for parts of the diagram:

thus *acd* can mean
{all elements not in *A only*}.

* * *

You have met enough of Set Algebra to consider it a study as interesting, or as boring, at least as useful, or as useless, as the Number Algebra of your early School Mathematics. You have seen certain ways in which the Algebras are similar in *structure* and in others dissimilar. $a . (x + y) = a . x + a . y$ is similar to $A \cap (X \cup Y) = (A \cap X) \cup (A \cap Y)$ by interchanging $\cup, \cap$ with $+, .$ but $A \cap A = A$ is not comparable with $a . a = a^2$.

In each of these Algebras some idea of the meaning of the symbols was expected at the beginning. Numbers and Sets lead to the symbols which later exist by themselves without constant interpretation as numbers or sets, and the various operators have verbally expressible meanings. But what of an Algebra presenting a *structure* though no interpretation? We would have, so to speak, Pure Algebra, or *Abstract* Algebra. Such a formulation is known as a Boolean Algebra, after the English mathematician George Boole whose book, *An introduction to the Laws of thought*, contained, in 1854, the first treatment of such an idea.

A complete self-consistent Boolean Algebra can be given by:

> *Symbols* A, B, C . . . (without meaning)
> *Operators* $\cup$ and $\cap$ (without meaning)
> *Rules* (1) Both operators are Commutative.
> (2) Each operator Distributes over the other.
> (3) Symbols 0 and 1 exist defined by $A \cup 0 = A$ and $A \cap 1 = A$.
> (4) Inverses exist defined by $A \cup A' = 1$ and $A \cap A' = 0$.

You can see, perhaps, that Set Algebra is a *model* of this Boolean Algebra, this purely axiomatic Algebra. All Set Algebra can be deduced from four such rules. They define a closed and consistent symbolic system.

8. Algebras:
(d) Some Casual Uses of Sets

You MAY have found a lot of this Algebra too academic. Boolean Algebra is, perhaps, an esoteric delight! Set language, however, can be helpful in many ways without any precise algebraic formulation and some examples of this follow:

I. Problems soluble by Venn diagrams.

A young reporter was sent to find out how the people of a small town spent Sunday. He returned with the news that everyone went to Church, played Golf or had a Picnic! He justified this by quoting that, of the people he questioned,

> 28 per cent went to Church
> 22 per cent played Golf
> 20 per cent had a Picnic
> 10 per cent went to Church and played Golf
> 3 per cent played Golf and had a Picnic
> 16 per cent had a Picnic and went to Church

and 1 per cent went to Church, played Golf and had a Picnic on that particular Sunday. There is the 100 per cent!

Let us put this information into a Venn diagram:

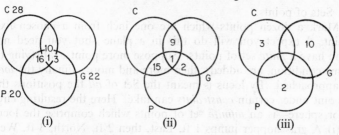

(i)   (ii)   (iii)

It is clear by subtraction, how many elements are in certain other Sets, fig. ii, and then fig. iii. By addition, the total percentage is only 42, not 100! Only 42 per cent of people questioned had gone to Church, played Golf or had a Picnic.

Suppose $n \cdot A$ means the *number* of members in a Set A. Then the reporter's findings can be summarised as:

$n \cdot C = 28$ $n \cdot C \cap G = 10$ $n \cdot C \cap G \cap P = 1$
$n \cdot G = 22$ $n \cdot G \cap P = 3$ where, in each case, the
$n \cdot P = 20$ $n \cdot P \cap C = 16$ unit is per cent.
—
70   29

Referring back to fig. i–iii we see the calculation exemplifies that

$n \cdot C \cup G \cup P = n \cdot C + n \cdot G + n \cdot P$
                              70
$- n \cdot C \cap G - n \cdot G \cap P - n \cdot P \cap C + n \cdot C \cap G \cap P$
$-$ 29 $+ 1 = 42.$

This formula is a general one for three Sets. Check its validity by considering a diagram labelled with small letters for the several parts, as in the last chapter.

Try these questions:

(i) Of 100 housewives, the same number were found to read just one of each of three magazines, whilst a quarter of them read all three. Some read only two; 15 read *Woman* and *Woman's Own*, 8 *Woman's Own* and *Woman's Realm*, 10 *Woman's Realm* and *Woman*. How many of them read *Woman*?

(ii) To become a House Prefect a boy must play in one, at least, of a School's 1st teams, the Rugby XV, the Cricket XI or Tennis VI. 3 boys play *only* rugger and cricket, 4 *only* rugger and tennis and 1 boy *only* cricket and tennis. How many Prefects are there?

II. Sets of points.

Mark a dozen points which are one inch from a chosen fixed point. I expect you will do this in a plane—but you need not! You have a *finite* set of points. Suppose more points, satisfying the same condition, are added: then more and more. Finally the *locus* is approached. By locus is meant the Set of *all* the positions that a point under certain *constraints* can take. Here the resulting circle —or sphere—is an *infinite* set of points which comprise the locus.

(i) A grasshopper jumps 1 ft. East, then 2 ft. North, 3 ft. West, 3 ft. South and 1 ft. East. Draw the locus as it is on the ground, i.e. a finite set of positions taken by the insect.

(ii) On squared paper draw the line $x + y = 2$ and call it L. This line is an infinite set of points. Draw also the line M, $y = x$.

What is the Set $L \cap M$?

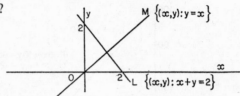

(iii) Use squared paper to draw a *lattice:* mark, for instance, only points with integral co-ordinates, as in fig. i. Draw any straight line

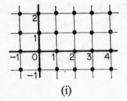   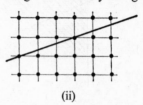

(i)   (ii)

through two or more of the lattice-points, as in fig. ii. *Three* sets of lattice-points are specified, those above the line, those on the line and those below the line.

III. Linear Programming.

This last example of point-sets determined by a straight line, leads to possible solutions of problems depending upon proportionate or part-proportionate variables under *constraint*.

A farmer wishes to sow crops to yield as high a profit as possible.

Facts	Symbols
He has two possible crops	x acres of A, y of B
I. 100 acres available for sowing	$x + y \leqslant 100$
II. Labour: A needs 1 man-day per acre, B 3 man-days. 150 man-days available	$x + 3y \leqslant 150$
III. He has £600 for seed: £5 per acre A, £10 B	$5x + 10y \leqslant 600$
Anticipated yield: £20 per acre A, £50 B	$20x + 50y$ to be *maximum*

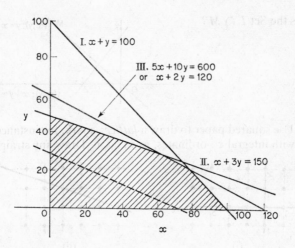

Graphs I, II and III determine the *domain* of the solution. Obviously x and y must be positive.

The Set of points which constitute possible answers to the farmer's problem are within the shaded area—just as in a Venn diagram. But $20x + 50y$ must be a maximum. So the solution point lies on the line $20x + 50y = $ max. or $2x + 5y = c$, say. One such line is shown dotted. Move a ruler parallel to this line to find the maximum, integral, point in the Set. The solution seems to be (60, 30). The man should plant 60 acres of crop A and 30 of B. Note that although this takes the full £600 and employs all the man-power, only 90 acres are planted.

Try this: 40 lb. of coffee and 15 lb. of chicory are to be mixed to produce two brands, one in the ratio 5:1 and the other 2:3. If the profit on the first is half as much again as that on the second, what is the most profitable number of pounds of each mixture to be made?

9. Algebras: (e) Logic

IN SPAIN an *Algebrista* is not someone, like you, trying to do some Algebra, but a Bone-setter! In English there are examples in the sixteenth and early seventeenth centuries of the word *algebra* being used for the treatment of fractures, for the 'putting together of broken parts'. And this is the meaning of the Arabic *al-jebr* from which the word comes. The Arab mathematician Abu Abdallah Mohammed Ibn Musa al-Khowarizmi compiled a treatise on Hindu calculation which carried forward the Arabic system of numeration and, in Baghdad about 830 AD, a work entitled 'al-jebr wa mukabala'—*restoration and reduction*—which dealt with the solving of certain practical problems by the use of symbols and equations, the *restoration* referring to the correct manipulation of the equations and the *reduction* to their simplification.

Algebra, in the sense of the use of symbols, came into Europe from this book and to England, via Italy, by the sixteenth century. The generalised use of Symbols, of course, is much more recent; the investigation into algebraic structure belongs to the nineteenth century and later.

We have glanced at the use of symbols standing for statements and sentences, for numbers and sets, and symbols standing just for themselves, and seen how their behaviours compare. Quite distinct interpretations and operations may show similar patterns. Implication Algebra may be expressible in Set terms. The sentences:

> All mathematicians are clever.
> All who cannot read time-tables are stupid.

may be expressed as:

$$M \Rightarrow C \text{ or as } M \subset C \text{ or}$$
$$T' \Rightarrow C' \qquad C \subset T$$
$$\text{so } M \Rightarrow T \text{ or} \qquad M \subset T$$

33

thus: All mathematicians can read time-tables
or: The set of mathematicians is a sub-set of the set of those who
can read time-tables. The sentences:

> All mathematicians are clever.
> Some mathematicians are agreeable.

have no unique conclusion. We have $M \subset C$, mathematicians are
a sub-set of clever people; and $M \cap A \neq \emptyset$, there are some members
common to the set of mathematicians and to the set of agreeable
people. The Venn diagram gives:

either: Some clever people are agreeable,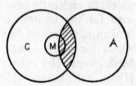
or: Some agreeable people are clever.

The four types of sentence in traditional logic are easily shown by
Venn diagrams.

Universal Affirmative.
All boys are good.

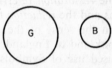

Universal Negative.
No boy is good.

Particular Affirmative.
Some boys are good.

Particular Negative.
Some boys are not good.

Such sentences express *Propositions*. A proposition is the *meaning* of
an indicative sentence. Sentences which are questions, commands or
exclamations do not give form to propositions.

Use Sets to determine whether the following *Syllogisms* are *valid:*

(i) It always rains on Sunday.

<u>Today it is raining</u>

Today is Sunday

(ii) All animals with pouches are marsupials.

<u>A kangaroo has a pouch</u>

A kangaroo is a marsupial

A syllogism consists of three propositions, each containing two *terms*, of the form

$$\begin{array}{cc} M \text{ is } P & \text{or} \quad P \text{ is } M \\ S \text{ is } M & \text{or} \quad M \text{ is } S \\ \hline & S \text{ is } P \end{array}$$

where M is the *middle term*, the repeated term, and S, P are the *subject* and the *predicate*.

Example (i) above	Example (ii) above
P is M	M is P
S is M	S is M
S is P	S is P
is not valid	is valid

There are four M, P, S patterns of syllogism and each of M, P, S may be any one of the four types of proposition already mentioned. Thus in all there are 256 schemes of syllogism! In the six books of the 'Organon' Aristotle developed a study of logical reasoning based on a classification of the Syllogism. Only 19 of the 256 have valid conclusions. Of these, some are 'perfect', some 'imperfect'. Whereas, now, the manipulation of Logic may depend upon an algebraic formulation, the classical Aristotelean Logic depended upon the reduction of an imperfect syllogism to a perfect one. The forms of reduction were memorised, in Mediaeval times, by Latin mnemonics. The substance and classification of logical reasoning was entirely verbal and remained so for centuries. The simplification and extension of the understanding of logical processes by the use of symbols was a fairly recent development. Its history starts, perhaps, with Leibnitz in the later seventeenth century and continues, with Euler, through the eighteenth: in this country, during the nineteenth century, William Hamilton, professor of logic at Edinburgh, Augustus de Morgan, professor of mathematics at

University College, London, George Boole, self-taught son of a Lincoln shoemaker who became professor of mathematics at Cork, and John Venn of Cambridge reached beyond the restrictions of classical logic to achieve greater generality in deductive reasoning through the use of symbolic processes. Boole's aim was '. . . to investigate the fundamental laws of the mind by which reasoning is performed; to give expression to them in the symbolical language of a Calculus, and upon this foundation to establish the science of Logic . . .'

Sentence Logic is a *model* of a Boolean Algebra.

> *Symbols* p, q, r . . . for propositions.
> *Operators* $\lor$ stands for the *connective* word, 'or'
> $\qquad\qquad \land$ stands for the connective, 'and'
>
> These are the *disjunction* and the *conjunction*.
>
> *Rules*   I. Commutative $p \lor q = q \lor p$ $p \land q = q \land p$
>   II. Distributive $p \lor (q \land r) = (p \lor q) \land (p \lor r)$
> $p \land (q \lor r) = (p \land q) \lor (p \land r)$
>
>   III. *Truth values* 0 and 1 signifying a universally false and a universally true sentence, satisfy
>
> $$p \lor 0 = p \qquad p \land 1 = p$$
>
>   IV. Negation, *not-p* is given by
>
> $$p \lor p' = 1 \qquad p \land p' = 0$$

Check, by taking particular sentences for p, q, r, or by putting the rules into words, that they seem reasonable. From propositions manipulated by these rules may be deduced all the apparatus of the Algebra: every true sentence is provable. Here are some expressions: verify that they are justified and specify the *dual* of each, obtained by interchanging the operators and the 0 and 1 where they occur.

     (i) $p \lor p = p$ (vi) $p \land q' = q' \land p$
     (ii) $p \land (p \lor q) = p$ (vii) $p \land q \land q' = 0$
     (iii) $p \land 0 = 0$ (viii) $(p \land q)' = p' \lor q'$
     (iv) $p \lor 1 = 1$ (ix) $p \land (q \land r')' = (p \land q') \lor (p \land r)$
     (v) $p = (p')'$ (x) $(p' \lor q)' \lor p = p$

Try to prove the last four from the given rules!

The composite sentences made up with the connectives are, in effect, being classified as either true or false. This suggests a comparison with Set Algebra that is concerned with classification. Look back to the examples (i) to (ix) of Chapter 7 and reconsider these in the symbols of Sentence Logic used here.

Sentence Logic, based upon the Boolean axioms I–IV is concerned with the conditions under which a complex of sentences can be said to be *true*. The truth depends on the validity of the developed connections between the originating propositions. It has nothing to do with the meaning that is any particular proposition, only with the correctness of the interrelation of the propositions in a composite sentence. It is not concerned with the interpretation of the symbols but with their valid relations through the connectives *and, or, not,* and *implies* in its specified sense.

The propositional symbols p, q, r . . . can only take the two truth values 0 and 1. If a mathematical expression is justified for *every* value possible of the variables it contains, it may be said to be *true*. (An algebraic expression can only be tested in this way if the number of possible values of the variables is *finite*: it can achieve nothing where, as in number algebra, there is an unlimited choice of values for the variables.) If a sentence, however complicated, is valid for values 0 and 1 of each of the component symbols, it must be logically justified.

The validity, then, of an expression in Sentence Logic, may be tested by a *Truth Table*. A sentence is *valid* if true for the possible values T or 1 for *true*, and F or 0 for *false*. A single proposition can have only one of two values, T or F. The Truth Table for p' is therefore:

p	p'
T	F
F	T

A Table for p, q and some connectives will require 4 lines as each of p, q may be T or F: one for p, q, r will need 8 lines, and so on. Here are two examples of such validity tests, the first to show $p \wedge 0 = 0$ and the second, an example of de Morgan's law, the dual of (viii) above.

O	p	$O \wedge p$		p	q	$p \vee q$	$(p \vee q)'$		p'	q'	$p' \wedge q'$
F	T	F		T	T	T	F		F	F	F
F	F	F		T	F	T	F		F	T	F
				F	T	T	F		T	F	F
				F	F	F	T		T	T	T

In both of these examples we see the derived underlined columns are identical and that the relations are thus shown to be valid.

Test some of the relations you have already met, in this way.

Such tests will prove whether the symbolically composed sentences are obtainable from the original propositions. A Truth Table will show whether a derived sentence is valid, contradictory or ambiguous. It must be stressed we are here dealing *only* with propositions which can be labelled singly as clearly possessing one of two possible truth values.

10. Algebras: (f) Matrix Algebra*

Numbers are put into a square or rectangular mould which we shall represent by the use of ( ) brackets. Each of the following is a *Matrix*.

$$\begin{pmatrix} 1 \\ 2 \end{pmatrix} \qquad \begin{pmatrix} 3 & \frac{1}{2} & 1 \\ 0 & 2 & 2 \\ \cdot 7 & 4 & 6 \end{pmatrix} \qquad \begin{pmatrix} -1 & 7 \\ -4 & 3 \end{pmatrix} \qquad \begin{pmatrix} a & b & c \\ x & y & z \end{pmatrix}$$

* To the reader: this and the subsequent chapter may prove rather hard and unpalatable for those who do not like messing about with algebraic symbols.

Let us try to build an Algebra of 2×2 *matrices* like the third example above. Capitals A, B, C, . . . will stand for such arrangements of numbers but we are not going to give any *meaning* to such an array. 'Multiplication' of two matrices shall be defined as the sum of the products of corresponding elements of rows of the left matrix and columns of the right matrix. An example will make this clearer. Suppose

$$A = \begin{pmatrix} 1 & 2 \\ 3 & 4 \end{pmatrix} \quad \text{and } B = \begin{pmatrix} 7 & 5 \\ 6 & 0 \end{pmatrix}$$

then
$$AB = \begin{pmatrix} 1 & 2 \\ 3 & 4 \end{pmatrix}\begin{pmatrix} 7 & 5 \\ 6 & 0 \end{pmatrix} = \begin{pmatrix} 1.7 + 2.6 & 1.5 + 2.0 \\ 3.7 + 4.6 & 3.5 + 4.0 \end{pmatrix}$$

$$= \begin{pmatrix} 7 + 12 & 5 + 0 \\ 21 + 24 & 15 + 0 \end{pmatrix} = \begin{pmatrix} 19 & 5 \\ 45 & 15 \end{pmatrix}$$

where the *top* row of the product derives from movement *through* the *top* row of A and *down* the columns of B, and the *bottom* row derives from movement through the *bottom* row of A down the columns of B. Here is another example:

If $X = \begin{pmatrix} 0 & -3 \\ 1 & 2 \end{pmatrix}$ and $Y = \begin{pmatrix} -1 & 1 \\ 2 & 5 \end{pmatrix}$

then $YX = \begin{pmatrix} -1 & 1 \\ 2 & 5 \end{pmatrix}\begin{pmatrix} 0 & -3 \\ 1 & 2 \end{pmatrix} = \begin{pmatrix} 0 + 1 & +3 + 2 \\ 0 + 5 & -6 + 10 \end{pmatrix} = \begin{pmatrix} 1 & 5 \\ 5 & 4 \end{pmatrix}$

Practice the method many times until it becomes automatic. Choose any two matrices P and Q, and work out both PQ and QP. You will almost certainly find $PQ \neq QP$. The first Rule of the Algebra, then, is that, in general, it is *not* commutative. We must refer to left-multiplication and right-multiplication, or pre- and post-multiplication, the written order of the letters signifying which is meant.

Some of the natural numbers 1, 2, 3 . . . have properties, under multiplication, which are immediately evident, $2.0 = 0$, $3.0 = 0$ . . . $2.1 = 2$, $3.1 = 3$, $4.1 = 4$ . . . $22 = 2.11$, $333 = 3.111$ . . . A matrix can be made with the natural numbers. Are there matrices which behave like 0, or like 1, or which multiply each digit? Make up some arrays which *might*, you think, behave in some particular way and try them under multiplication with any other matrix.

You are sure to find that $\begin{pmatrix} 0 & 0 \\ 0 & 0 \end{pmatrix}$ annihilates any matrix. This is the *zero* matrix Z such that $AZ = ZA = Z$. $\begin{pmatrix} 1 & 1 \\ 1 & 1 \end{pmatrix}$, $\begin{pmatrix} 2 & 2 \\ 2 & 2 \end{pmatrix}$ and so on, don't appear to have special significance, but if any matrix is multiplied by the second, all the elements of the product are double those when it is multiplied by the first; or three times, and so on. This suggests

$$\begin{pmatrix} 2 & 2 \\ 2 & 2 \end{pmatrix} = 2\begin{pmatrix} 1 & 1 \\ 1 & 1 \end{pmatrix} \text{ and } \begin{pmatrix} 3 & 3 \\ 3 & 3 \end{pmatrix} = 3\begin{pmatrix} 1 & 1 \\ 1 & 1 \end{pmatrix}$$

and so on, thus defining scalar multiplication by some such rule as $k(a) = (ka)$. Consider this for yourself.

Now $\begin{pmatrix} a & b \\ c & d \end{pmatrix}\begin{pmatrix} 1 & 1 \\ 1 & 1 \end{pmatrix} = \begin{pmatrix} a + b & a + b \\ c + d & c + d \end{pmatrix}$

and $\begin{pmatrix} 1 & 1 \\ 1 & 1 \end{pmatrix}\begin{pmatrix} a & b \\ c & d \end{pmatrix} = \begin{pmatrix} a + c & b + d \\ a + c & b + d \end{pmatrix}$.

Each product is $\begin{pmatrix} a & b \\ c & d \end{pmatrix}$ with the addition of a term to each element. Can these extra terms be removed so as to leave the $\begin{pmatrix} a & b \\ c & d \end{pmatrix}$ started with? $\begin{pmatrix} 1 & 0 \\ 0 & 1 \end{pmatrix}$ will do this. It is the *Unit* matrix I such that $AI = IA = A$.

If a is a natural number, is there a number such that $aa' = 1$? If $a = 2$ then $a' = \frac{1}{2}$, the reciprocal of 2, not a natural number but a fraction. If A is a matrix, is there an A' such that $AA' = I$? Suppose

$$A = \begin{pmatrix} 2 & 3 \\ 1 & 2 \end{pmatrix} \text{ Can you find } a, b, c, d,$$

so that $\begin{pmatrix} 2 & 3 \\ 1 & 2 \end{pmatrix}\begin{pmatrix} a & b \\ c & d \end{pmatrix} = \begin{pmatrix} 1 & 0 \\ 0 & 1 \end{pmatrix}$?

Yoy may try intelligent guessing and checking: you may compare the product with I and find the numbers from simultaneous equations; or you can be lazy, and just verify that a, b, c, d are 2, −3, −1, 2. A pattern is suggested:

$$\begin{pmatrix} 2 & 3 \\ 1 & 2 \end{pmatrix} \text{ leads to } \begin{pmatrix} 2 & -3 \\ -1 & 2 \end{pmatrix}$$

Try this pattern for $\begin{pmatrix} 2 & 1 \\ 5 & 3 \end{pmatrix}$

You discover $\begin{pmatrix} 2 & -1 \\ -5 & 3 \end{pmatrix}$ does not lead to I but that $\begin{pmatrix} 3 & -1 \\ -5 & 2 \end{pmatrix}$ does.

Both post- and pre-multiplication of $\begin{pmatrix} a & b \\ c & d \end{pmatrix}$ by $\begin{pmatrix} d & -b \\ -c & a \end{pmatrix}$ give I, but multiplied by the *number* $ad - bc$. We shall represent this number associated with a matrix A by $|A|$. Finally we have

$$AA' = A'A = |A|I.$$

Satisfy yourself that this particular multiplication *is* commutative. Practise the finding of A' until it is understood. What happens if $|A| = 0$?

Find some examples of $AB = Z$. Does $AB = Z$ require $A = Z$ or $B = Z$? What can you say about the equation $AB = AC$?

The Algebra, so far, can now be formulated.

Symbols $A, B, C \ldots$ for 2×2 matrices.

Operator 'Multiplication'.

Rules (i) Associative (check this for yourself), but *non-*Commutative, in general.

    (ii) Scalar multiplication $k(a) = (ka)$

    (iii) $Z = \begin{pmatrix} 0 & 0 \\ 0 & 0 \end{pmatrix}$ such that $AZ = ZA = Z$

    (iv) $I = \begin{pmatrix} 1 & 0 \\ 0 & 1 \end{pmatrix}$ such that $AI = IA = A$

    (v) Matrices A' exist such that

$$AA' = A'A = |A|I.$$

11. Algebras:
(g) Some Applications of Matrices

A MATRIX is a display of information, any information which can be shown by a rectangular array of numbers. A bus time-table is a matrix; a log table is a matrix; a price-list is a matrix, but if the Algebra compiled in the last chapter is to be applied to a matrix which *now* is to have meaning, the results of the operation of combination must have meaning too. Multiplication seems hardly likely to give useful results if tried on time-tables, logs or price-lists, although addition might! It can, however, be used on any two matrices where the number of columns of the left matrix equals the number of rows of the right one. So:

if $A = \begin{pmatrix} 1 & 0 \\ 0 & 1 \\ 1 & 0 \end{pmatrix}$ and $B = \begin{pmatrix} 1 & 0 & 1 \\ 1 & 0 & 1 \end{pmatrix}$ then $AB = \begin{pmatrix} 1 & 0 & 1 \\ 1 & 0 & 1 \\ 1 & 0 & 1 \end{pmatrix}$ and

$$BA = \begin{pmatrix} 2 & 0 \\ 2 & 0 \end{pmatrix}$$

We shall confine ourselves, mainly, nonetheless, to the 2 × 2 arrays already met.

The older School Algebras proudly—and very usefully—presented pages and pages of exercises, on the fine principle that 'practice makes perfect'. Suppose 100 examples of the following type were required:

Solve the equations
$$\begin{aligned} 3x + 2y &= 8 \\ x + y &= 3 \end{aligned}$$

Provided every pair of equations was in this form, each example could be expressed as a matrix; this as

$$\begin{pmatrix} 3 & 2 & 8 \\ 1 & 1 & 3 \end{pmatrix} \text{ and the solution set } \{2, 1\} \text{ as the matrix } \begin{pmatrix} 2 \\ 1 \end{pmatrix}$$

All the others would be written as simply. If you object to the loss of x and y—a reasonable objection if they are *variables* rather than unknowns—then the example given, would be

Solve $\begin{pmatrix} 3 & 2 \\ 1 & 1 \end{pmatrix} \begin{pmatrix} x \\ y \end{pmatrix} = \begin{pmatrix} 8 \\ 3 \end{pmatrix}$ and the solution $\begin{pmatrix} x \\ y \end{pmatrix} = \begin{pmatrix} 2 \\ 1 \end{pmatrix}$

Check for yourself that multiplication of the LHS leads to the original form of the equations.

Look again at the last version and write it $AX = B$.

Now	$AX = B$	
so	$A'AX = A'B$	mult. by adjoint A
or	$\lvert A \rvert IX = A'B$	by rule (v)
i.e.	$\lvert A \rvert X = A'B$	by rule (iv)
	$X = A'B/\lvert A \rvert$	

The equations are thus solved by a *general* method. In particular,

$$\begin{pmatrix} 3 & 2 \\ 1 & 1 \end{pmatrix} \begin{pmatrix} x \\ y \end{pmatrix} = \begin{pmatrix} 8 \\ 3 \end{pmatrix}$$

$$\Rightarrow \quad \begin{pmatrix} x \\ y \end{pmatrix} = \begin{pmatrix} 1 & -2 \\ -1 & 3 \end{pmatrix} \begin{pmatrix} 8 \\ 3 \end{pmatrix} = \begin{pmatrix} 2 \\ 1 \end{pmatrix} \qquad \text{Done!}$$

Here is another,

$$\begin{pmatrix} 1 \cdot 6 & - \cdot 4 \\ - \cdot 5 & 2 \cdot 0 \end{pmatrix} \begin{pmatrix} p \\ q \end{pmatrix} = \begin{pmatrix} 1 \\ 2 \end{pmatrix}$$

$$\Rightarrow \quad 3 \begin{pmatrix} p \\ q \end{pmatrix} = \begin{pmatrix} 2 \cdot 0 & \cdot 4 \\ \cdot 5 & 1 \cdot 6 \end{pmatrix} \begin{pmatrix} 1 \\ 2 \end{pmatrix} = \begin{pmatrix} 2 \cdot 8 \\ 3 \cdot 7 \end{pmatrix}$$

$$\begin{pmatrix} p \\ q \end{pmatrix} = \begin{pmatrix} \cdot 93 \\ 1 \cdot 2 \end{pmatrix} \text{ to 2 sig. fig.}$$

Write out pairs of linear equations of your own, and solve them.

The method can be applied to a set of *any number* of linear equations $AX = B$ where A is a square matrix containing the coefficients of x, y, z . . ., X a column matrix x, y, z . . . and B a column matrix of numbers. The finding of A' and the multiplication is much more arduous but the general method is the same, $\lvert A \rvert X = A'B$. The actual manipulation is well suited to adaptation for machine and computer calculation.

Not only do the matrices provide a general method for dealing with sets of linear equations, however; they also provide a way of

judging the nature of a set of such equations. Suppose four quantities
x, y, z, t are connected by four relations such as $1 \cdot 21x - 4 \cdot 2y + 8 \cdot 17z - t = 27 \cdot 3$. Is there a set of *unique* values $\{x, y, z, t\}$
which satisfy the equation? A matrix will provide the frame of an
answer.

This is too involved a matter to investigate here, but if you are
interested enough, consider a simpler case of pairs of straight lines
in a plane—drawn on graph paper. Consider:

   (i) $x - y = 0$ and $x + y = 2$ meet at (1, 1).
  (ii) $x + y = 1$ and $x + y = 2$ do not *meet* but are parallel.
 (iii) $x + y = 1$ and $2x + 2y = 2$ are the same line.

In matrix language these pairs could be expressed as

$$\begin{pmatrix} 1 & -1 & 0 \\ 1 & 1 & 2 \end{pmatrix} \text{ and } \begin{pmatrix} 1 & 1 & 1 \\ 1 & 1 & 2 \end{pmatrix} \text{ and } \begin{pmatrix} 1 & 1 & 1 \\ 2 & 2 & 2 \end{pmatrix}$$

Try to find out in what ways these arrays show that the equations
(i) have a unique solution, (ii) have no solution—are inconsistent,
(iii) have an infinity of solutions—are not independent.

*   *   *

You all know how a position may be fixed by numbers. A point
on a line may be determined by a single number. P is $3 \cdot 5$.

A point in a plane requires two numbers because of two *degrees of*

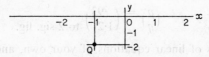

freedom. Q is $(-1, -2)$. A point in 3 dimensions needs three numbers. R is (1, 2, 3). How can a position be changed, algebraically?

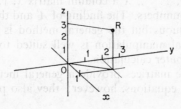

Clearly, by a change in these numbers, these co-ordinates. Let a point (x, y) in a plane be transformed to, or *mapped onto*, a point (X, Y) by relations

$$X = ax + by \text{ or } \begin{pmatrix} X \\ Y \end{pmatrix} = \begin{pmatrix} a & b \\ c & d \end{pmatrix}\begin{pmatrix} x \\ y \end{pmatrix} \text{ or } U = AV \text{ say.}$$
$$Y = cx + dy$$

The matrix A provides the mapping. A further transformation designated by a matrix B will provide a further mapping W where $W = BU = BAV$, and so on.

On squared paper draw the figure $ABCD$ shown below.

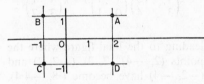

Use $\frac{1}{2}$ inch scale, and place 0 centrally on the paper. Apply the following matrix operations to A, B, C, D both graphically and algebraically,

in the order given: firstly $\begin{pmatrix} 2 & 0 \\ 0 & 2 \end{pmatrix}$, then $\begin{pmatrix} -1 & 0 \\ 0 & 1 \end{pmatrix}$ and then $\begin{pmatrix} 0 & 1 \\ 1 & 0 \end{pmatrix}$. The rectangle has become:

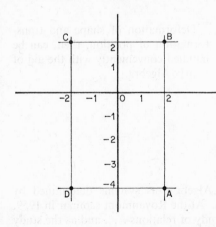

You have (i) doubled the lengths, (ii) reflected the rectangle in the y-axis and (iii) rotated it about the line

$$y = x. \quad A \ (2, 1)$$

or $\begin{pmatrix} 2 \\ 1 \end{pmatrix}$ became first

$$\begin{pmatrix} 2 & 0 \\ 0 & 2 \end{pmatrix}\begin{pmatrix} 2 \\ 1 \end{pmatrix} = \begin{pmatrix} 4 \\ 2 \end{pmatrix} \text{ then}$$

$$\begin{pmatrix} -1 & 0 \\ 0 & 1 \end{pmatrix}\begin{pmatrix} 4 \\ 2 \end{pmatrix} = \begin{pmatrix} -4 \\ 2 \end{pmatrix}$$

and finally $\begin{pmatrix} 0 & 1 \\ 1 & 0 \end{pmatrix}\begin{pmatrix} -4 \\ 2 \end{pmatrix} = \begin{pmatrix} 2 \\ -4 \end{pmatrix}$ or $(2, -4)$; and similarly for B, C, D.

If you have managed this satisfactorily, try these further mappings,

$$\begin{pmatrix} 1 & 1 \\ 0 & 1 \end{pmatrix} \text{ and } \begin{pmatrix} \cos\alpha & -\sin\alpha \\ \sin\alpha & \cos\alpha \end{pmatrix}$$

where α is the smallest angle of a 3, 4, 5 triangle.

The result of the first three changes were produced by

$$\begin{pmatrix} 0 & 1 \\ 1 & 0 \end{pmatrix}\begin{pmatrix} -1 & 0 \\ 0 & 1 \end{pmatrix}\begin{pmatrix} 2 & 0 \\ 0 & 2 \end{pmatrix} = \begin{pmatrix} 0 & 1 \\ 1 & 0 \end{pmatrix}\begin{pmatrix} -2 & 0 \\ 0 & 2 \end{pmatrix} = \begin{pmatrix} 0 & 2 \\ -2 & 0 \end{pmatrix}$$

Check this graphically. The other two were produced by

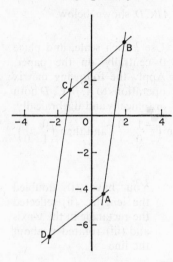

$$\tfrac{1}{5}\begin{pmatrix} 4 & -3 \\ 3 & 4 \end{pmatrix}\begin{pmatrix} 1 & 1 \\ 0 & 1 \end{pmatrix} = \tfrac{1}{5}\begin{pmatrix} 4 & 1 \\ 3 & 7 \end{pmatrix}$$

leading to the final figure where the points $(2, -4)$, $(2, 2)$, $(-2, 2)$ and $(-2, -4)$ have become $(\cdot8, -4\cdot4)$, $(2, 4)$, $(-1\cdot2, 1\cdot6)$ and $(-2\cdot4, -6\cdot8)$. $ABCD$ is now a parallelogram.

Deformation of shape and transformation of position, then, can be handled conveniently with the aid of matrix Algebra.

*   *   *

We have been looking at Algebras as systems determined by Symbols, Operators and Rules. At the Royaumont seminar in 1959, Algebra was defined, 'as the study of relations . . . and as the study of operations on given sets which create new entities . . .' Perhaps the section of work you have just completed shows a little of the power the algebraic expression of an operation may have.

12. What is Geometry?

WHAT IS BIOLOGY? The Greek etymology of the word gives the answer, βίος life, λογία study; the study of life. What is Geometry? γῆ Earth, μετρία measuring. Biology is a fairly recent word, Geometry an old one; even so, Earth measuring suggests Surveying not Geometry. In the North Choir aisle of Bristol Cathedral there is a stone slab from a twelfth century tomb whose inscription reads, 'Willaim le geometer'. William, presumably, was concerned in some way with the laying out of the ground and the building of the original Abbey. After the Great Fire of 1666 Evelyn wrote of 'geometers' at work in the City of London.

The Egyptians were admirable practical mathematicians. Their large scale mensuration work in the building of temples and pyramids could well be considered as Earth measuring; as Geometry or Surveying. Much of the factual knowledge that lies behind early Greek mathematics came from Egypt. And any hazy perception *you* may have of the nature of Geometry will have been formed, probably, by your remembering something of Pythagoras or Euclid!

Thales, who correctly predicted an Eclipse in 585 BC, collected many facts from the Egyptians and subsequently advised the young Pythagoras to visit that country. Thales was a merchant and a public figure as well as a Philosopher and Mathematician. His ultimate belief was that Water was the essence of all things, ἄριστον μὲν ὕδωρ, 'Water is best', a fact amusingly referred to by Walter Savage Landor in the lines,

> Ye lie, friend Pindar and friend Thales.
> Nothing better than Water? Ale is!

But his importance to us is that he first attempted the deductive proof of propositions. The Egyptians knew from practical experience that lengths related to 3, 4, 5 set out a right angle, that certain rectangles and triangles had ascertained areas and could be compared with others for computation, but they did not have the *general* facts

47

of which these are *particular* instances. A host of problems previously solved was the basis for adaptation for new problems. Thales managed to co-ordinate some of the accumulation of knowledge; to relate some parts to others and to justify the correctness of one assumption by reference to another.

Every diagram is a particular instance, a special case. To say, *ABC* is a triangle, is like saying, *x* is a number.

A study of shapes cannot be Geometry until a shape can be considered apart from what is inferred from a lot of instances. Thales started such abstraction.

Pythagoras, born when Thales was about 60, greatly continued this, extending the deductive chain of argument and generalising the particular into the abstract concept. Every schoolboy knows Pythagoras by one theorem! Even this is enough, however, to show the power of abstraction. Compare the use of 12 yards of rope to determine a corner of a wall, with the Converse of Pythagoras' theorem:

If, in a triangle, the area of the square on the longest side is equal to the sum of the areas of the squares on the other two sides, then the triangle is right angled opposite that side.

Not only is there generality independent of any diagram, but what is in effect, a definition of a *space*, a rigidity given to the behaviour of three lines.

The achievement of Pythagoras is vast and fascinating. Find out for yourself more of his Mathematics, divided, as it is, into Arithmetic, Geometry, Music and Astronomy—later the *Quadrivium* of Medieval learning—and about the beliefs and tenets of the Pythagoreans, as various as the transmigration of the soul, the mystic nature of numbers, and abstention from the eating of beans!

The world of abstraction generalised from the world of the particular, evidences the separate worlds of the physical and the geometrical. Geometry is concerned with shapes: a physical shape may be lovely to behold but there might be a perfection of form which can only be imagined, of a higher order of beauty: an *ideal* which the real might approach. To the Greeks beauty on Earth was a poor reflection of heavenly Beauty, the ἰδέα, idea of perfection, the archetype of which individual things are imperfect copies.

The study of shapes, then, might originate in the practical and particular, be inter-related by similarities, carried over to concepts tied by deductive reasoning and conclude by aiming at a definition of beauty as an ultimate perfection of form. But this sequence is not wholly academic or philosophical. As fact leads to fact and conclusion to conclusion, common sense mentally re-connects with the physical, and a greater understanding of the real world emerges.

The Library at Alexandria, finally holding some three-quarter million books, was started about 300 BC when Euclid was one of the teachers at the University there. His *Elements* attempted a classification of the Mathematics of the time. The XIII Books cover number theory, as well as the Geometry which was to dominate geometrical thinking for 2000 years and which, says Bertrand Russell, '. . . was still, when I was young, the sole acknowledged text-book of geometry for boys'.

Euclid's work was a brilliant compilation. In the geometrical books he built a rigorous deductive chain of reasoning connecting and inter-connecting the existing concepts of ideal shapes. The whole was the finest exercise and example of the Greek deductive method. Its summit was Book XIII dealing with the five 'Platonic' polyhedra, the only five completely *regular* plane-faced solids.

A *postulate* of parallel lines involves equal corresponding and alternate angles. This leads to the exterior angle property of a triangle, the angle sum of a triangle and on to the angles of polygons *or*, via isosceles triangles, to the angle properties of a circle. The

etc.   etc.

The 5 regular solids

Parallel lines

Cyclic quadrilaterals

above is not so much an instance from Euclid as an example of the method. The facts relating to the diagrams are ones you remember, I expect.

The regular solids are shown and named below. Convince yourself these 5 are the *only* regular polyhedra by considering the make

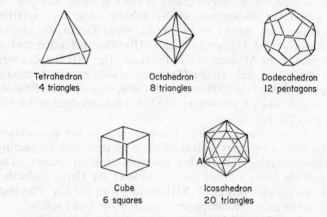

| Tetrahedron
4 triangles | Octahedron
8 triangles | Dodecahedron
12 pentagons |

| Cube
6 squares | Icosahedron
20 triangles |

(Diagram based upon *The World of Mathematics*, p. 582)

up of a vertex from a paper *net*. Thus, taking the icosahedron as an example, a vertex *A* derives from sticking together the sides *a*, *b* of a nest of 5 equilateral triangles. Remembering that the angle

of an equilateral triangle is 60°, of a square 90°, of a regular pentagon and hexagon, 108° and 120°, show that such a configuration can *only* be possible with 3 or 4 or 5 equilateral triangles, with 3 squares or 3 regular pentagons. Six such triangles or three such hexagons lie in a plane and more figures than this are bound to cause overlapping and so no possible solid.

The power of Euclid lies in the rigour of the deductive reasoning leading from conclusion to conclusion, but any *truth* depends upon the initial axioms. Things are only true by experience if the axioms are true by experience. The geometrical truth is one of valid reasoning alone. Euclid realised he could not prove the geometrical existence of parallels but had to postulate them in a way which says, in effect:

Through any point *P* not on a straight line *L* there can be drawn just one line in the *P* and *L* plane which does not meet *L*. The

acceptance of this axiom determines the geometrical space in which
the corpus of Euclidean Geometry is justified. It cannot be *proved*

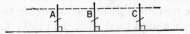

that *ABC* is a straight line without some such assumption. There
is inevitably a dichotomy between the logically acceptable and the
world as intuitively perceived.

Euclid was basic in Medieval learning, and later, although
Newton thought the Elements at first a 'trifling book', he was after-
wards humbly to recant! It was not until the parallel hypothesis
could be questioned—and ignored—that there could be a proper
separation of a geometrical world from the physical world.

The Geometry of Euclid is the investigation of space-relations
in a *plane*. The parallel postulate—or Pythagoras' theorem—will
determine a surface in which the sequential facts are valid. If a
plane is folded into a cylinder, say, or a cone—granted no distance
limitation—the same Geometry will be valid: facts about isosceles
triangles, circles, bisectors and so on, will be as in a plane. But a
straight line will not be as understood in a plane. It will be the
shortest line *in the surface*. I shall here use *strait* when referring to a
line in this sense.

A compilation of space-relation facts, logically connected,
could be attempted on a Globe, an Egg or a Saddle, and if properly
constituted the facts, though differing, would all be axiomatically
justified: the physical picture provided by the Globe, the Egg or
the Saddle, would be no more than the Blackboard as an aid to
Euclidean plane Geometry.

After some 2,000 years periodic attempts were made to prove
the axiom of parallels. The problem was finally dealt with in the
19th century by ignoring it. Consistent Geometries were evolved
in which there is no such premise. Consider the statement:

Through any point *P* not on a *strait* line *L* there can be drawn
no strait line which does not meet *L*.

Applied to a physical plane, it is nonsense: applied to an Euclidean
plane, it is false: but apply it to the Earth, to the surface of a sphere,
to an *elliptic* Geometry, and it is correct. Remember that a strait
line on the Earth is an arc of a Great Circle—there are any number

of lines through points *A* and *B* on a sphere, but only one strait line, i.e. the *geodesic* or shortest line on the surface. In such a Geometry the theorem of Pythagoras does not exist. Every pair of strait lines of the maximum finite length meet in two points, and the angles of a triangle always total more than 180°. Convince yourself, with a ball and a piece of string, if necessary, that these facts are reasonable when referred to a Globe.

The formulation of such a 2-dimensional Geometry does not, of course, specify a sphere, itself a 3-dimensional form or concept. Its axioms and development are such that the surface behaviour of shapes on a Globe provides a useful model.

In an elliptic Geometry there is *no* parallel to line *L* through point *P:* in parabolic Geometry—like Euclid's—there is just one: in hyperbolic Geometry there are many. The names of Gauss, Lobachevsky, Bolyai and Riemann are associated with the study of such Geometries in the last century. They emphasised the validity of non-Euclidean spaces and that the link of sense-perceived shapes with Euclidean space is not the only possible one. Riemannian Geometry led to new cosmological theories and was the frame within which Einstein was able to develop much of his work.

The mathematical world of a Geometry is justified by its axioms, not by any apparent reality. A physical sphere will aid the understanding of an elliptic Geometry, a blackboard, a parabolic, and a saddle, a hyperbolic Geometry. But even these realities have to be idealised in our minds if they are to help very much. We must be able to ignore the incidental physical limitations, the finite dimensions of the saddle or the edge of the blackboard, and equally, to be aware of the essential properties, the finite area of the sphere's surface or the never ending lengths of lines on a plane. We must feel confident that whatever happens in one place could happen in another. It is unthinkable that a carpet measured and cut to fit a room in Blackpool should not fit a room of the same measurements in Brighton! It is unthinkable that triangles *ABC* and *XYZ* should be *congruent*, i.e. superimposable by sliding, in one region of the surface and not in another. Clearly, then, the Egg mentioned earlier cannot be a model for a non-Euclidean Geometry—how awkward that equilateral triangles have varying angles according to their positions on the shell! This does not mean that there can be no Geometry of Eggs, but that the Geometry of such a shape must be able to contain the whole shape and not the surface only. The elliptic, parabolic and hyperbolic Geometries, already referred to, are plane in the sense that there is no dimension *outside* the surface. A small, blind, crawling insect which cannot fly or dig, living upon a walnut shell, cannot be aware of the undulation of his world, nor of its limitations. A 3-dimensional perception is needed to see how imprisoned he is! Perhaps a 4-dimensional perception is needed to see how imprisoned *we* are by the geometrical structure of our world!

An algebraical extension of dimension is essentially easy: the manipulation becomes more and more complex, however, and interpretation more difficult, artificial and terminological. A point in a line may be specified by a number, $+ x$ say, to the right and $-x$ to the left, of a fixed 0, so that $x^2 = r^2$ represents the set of point-positions or *locus*, such that a movable P may be a constant distance r from 0. In set language,

$$\{P(x):x^2 = r^2\}$$

In 2 dimensions we have the circle radius r,

$$\{P(x, y):x^2 + y^2 = r^2\}$$

In 3 dimensions, the sphere,

$$\{P(x, y, z):x^2 + y^2 + z^2 = r^2\}$$

In 4 dimensions, the hyper-sphere,

$$\{P(x, y, z, t):x^2 + y^2 + z^2 + t^2 = r^2\}$$

and so on.

A glance at an advanced school geometry text or a university book—or even, perhaps, some very new little book for primary schools—will suggest, I suspect, that Geometry is really Algebra, with a few diagrams thrown in to help the feeble! Algebraic geometry, it has often been said, begins with Descartes. René Descartes, soldier, philosopher and mathematician, published his 'Discours de la méthode pour bien conduire sa raison et chercher le vérité dans les sciences' in 1637. The third appendix, 'la Geometrie', is the source book of Cartesian Geometry. His work is both axiomatic and practical. He did not *invent* the association of algebra and geometry but carried it further than anyone before. The Cartesian co-ordinates that every schoolboy knows—the ordinary x, y of a graph—were his innovation. Position is determined by co-ordinates:

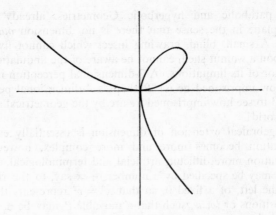

locus and shape by the set of positions a point may take. An equation determines a set of numbers which satisfy it. The locus or shape is the geometrical expression of the limitation imposed by the equation. Thus the equation $x^3 + y^3 = 3xy$ is the curve shown, because the infinite set of *real* values of x and y which satisfy the equation correspond to the infinite set of points which together form this Folium of Descartes. The equation $x^2 + y^2 = 0$ represents the single point-set $(0, 0)$ as $x = 0$, $y = 0$ are the only real numbers which satisfy the equation: alternatively, it is a circle of zero radius.

Descartes' own algebraic representation of such things was neither, however, as simple nor as complete as this suggests: he did not actually nominate a y-axis and restricted himself to positive values of x.

Only two generations after the *Discourse on Method*, Newton was powerfully investigating seventy-eight species of cubic curves by this Geometry, whereas, before Descartes, the vastly simpler Conic Sections were still being studied by methods essentially those of the Greeks nearly 2000 years earlier.

Algebra encourages a Geometry of movement, of a point tracing out a curve or taking up varying positions, rather than a study of static shapes. Some conception of a locus goes back to Thales, but Euclid is mainly concerned with the static and particular. The Greeks were much interested in movement but had not the equipment, nor perhaps the inclination, to investigate it geometrically. Congruency is, in a sense, an awareness of movement geometrically; a movement in a surface to superimpose one shape upon another: it is a transformation changing position but not shape. The Classical problems of Duplicating the Cube and Squaring the Circle are essentially kinetic, dealing, as they do, with the transformation of equi-spatial shapes. But Greek concern with the problems of movement is seen most clearly in Zeno's Paradoxes, of which Achilles and the Tortoise is the best known.

A direct concern with physical movement and physical perception has often thrust forward the advances in Geometry which itself has then provided the frame for further advances in physical understanding. Hero's observation of the reflection of light in 3rd century Egypt or Kepler's deductions from astonomical observations around 1600, increased knowledge of the physical world and confirmed later the nearness of reality to formulated geometrical patterns. Again, the attempts over centuries to achieve in drawing and painting a perspective which could be contrived by rule yet be satisfying to the eye and pleasing, meant geometrical searching for Leonardo da Vinci in Italy and Albrecht Dürer in Germany. This culminated in the compilation of a Projective Geometry by Gérard Desargues, an architect and engineer of the seventeenth century, a Geometry dealing with the properties of shapes under projection. In its long history this Geometry became a vehicle for expressing space-relation characteristics required in the theory of relativity.

Geometry is the investigation of space-relationships, independent of position, in a space defined by that Geometry. 'A game is what I say it is!', says the Geometer. But the realm is so large . . . and this is so limiting. What is Geometry? What a Geometer does. What does a Geometer do? Geometry!

13. Topology: (a) Euler

YOU ARE SURE to know the topography of the district in which you live. Topology and Topiary are two other words which derive from the Greek word for *place*! Our concern here is a look at a little Topology, the mathematics of the *essential* characteristics of shapes. Topologically, a circle and a triangle, a field and a pond, are all the same type of shape because the boundary of each encloses a certain area; a wedding-ring and a tea cup, similarly, are of the same type of shape, whereas an inner-tube, altho' ring-like, is a different shape, for the valve gives access to the 'inside'.

Many cheeses come to the table in wedge-shapes. Let us suppose the wedge has been so carefully cut that the faces of the piece are flat, meeting in straight lines and points. We are considering the cheese as a triangular prism. Clearly there are 5 faces, 6 vertices and 9 edges.

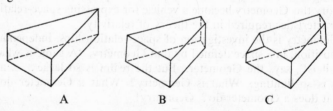

A   B   C

Such cheeses are usually cut, at table, with a knife and not dug into with a spoon or gouged out with a fork! Any cut will leave a piece which, of course, is smaller, but which is still a polyhedron—has plane faces. A further cut will do the same. The *essential* quality of the shape is that it remains polyhedral, perhaps as shown in the diagrams. Are there any other qualities which remain the same? Make a note of the number of faces (F), number of vertices (V) and the number of edges (E) of the three pieces of cheese and do likewise after further cuts of your own choosing, producing shapes X and Y, say. Do you notice any relationship between the F, V and E which stays constant despite the cuts?

Extend your list by writing in the numbers for solids you can visualise or happen to have available to handle. Here is a partially completed table such as you might draw up:

Name	F	V	E
Cheese A	5	6	9
Cheese B	6	8	12
Cheese C	7	10	15
Cheese X			
Cheese Y			
Oxo box			
Toblerone pkt.			
Square pyramid			
Threepenny piece	14	24	36
Dodecehedral paperweight	12	20	30
Pentagonal prism			
Hexagonal pyramid	7	7	12
. . .			

(It is worthwhile to compile the values for a series of prisms and pyramids.)

In each and every case

$$F + V = E + 2$$

This is Euler's relation between the numbers of faces, vertices and edges of a convex polyhedron. Certain modifications are necessary if solids have stellations, re-entries, 'gutters' or 'holes'. Its validity is actually much greater than the present approach suggests.

The correctness of the relation can be tested by starting with the 'simplest' solid and successively cutting off vertices. If a solid is bounded solely by plane surfaces the smallest possible number of faces is four. We then have a triangular pyramid or tetrahedron with $F = 4$, $V = 4$, $E = 6$ and these values satisfy the equation. Any cut which removes a vertex, adds three vertices, adds one face and adds three edges. Thus the left hand side of the equation is increased by three and so is the right hand side—the equation remaining satisfied: the relation still holds.

Leonhard Euler (1707–1783) is usually accorded the honour of
Father of Topology because of his address to the St. Peterberg
Academy in 1735 about the Königsberg puzzle, but the F, V, E,
relation for instance, was known to Descartes a century before.

Königsberg, now Kaliningrad, is built near the mouth of the river
Pregel and a part on the island of Kneiphof. In the eighteenth
century there were seven bridges as depicted. The puzzle was to

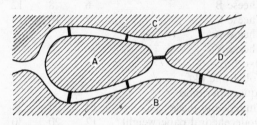

find if it were possible to walk over all the bridges in succession
without recrossing any one. Euler's investigation began with the
labelling of the map. The four land masses were *A*, *B*, *C*, and *D;*
the routes across the bridges were, for example, *AB* or *DCA*. In
effect he treated *A*, *B*, *C*, *D* as *points* so that
the map became a much simpler *network*, tho'
this second diagram is not actually in the
original paper. The *unicursal* nature of the
problem is now clearer. Can the pictures be

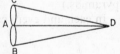

drawn in one continuous movement of the pencil? Try it! The second
diagram is topologically equivalent to the first as, for instance, the
London Underground map is topologically equivalent to the London
Underground system.

Try, also, to draw the following, each in one continuous movement:

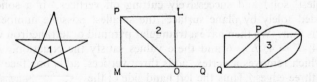

You find the Königsberg drawing impossible; it needs two distinct
strokes at least. Diagram 1 (above) can be described in one sweep,

starting and finishing at any chosen point: diagrams 2 and 3 prove possible only if the pencil moves from P to Q or Q to P.

Why should this be? How can the situation be known without tedious and uncertain testing by drawing? Euler proved four facts about the nature of a network. (Any configuration of lines you care to draw may be thought of as a network.) A network is compounded of branch-lines or *branches*. Each branch has two *ends* and connects *nodes* of the network. Thus in diagram 2, *MO OQ PL PO PM LO* are branches and P, L, M, O, Q are each nodes of the network. A node is *odd* or *even* according as an odd or even number of ends are at that node. P and Q are *odd nodes;* L, M, O are *even nodes*.

I. There is always an even number of odd nodes, or none.

II. A unicursal path from any point returning to that point, has no odd nodes.

III. A network with just two odd nodes may be traversed unicursally but only from one of those nodes to the other.

IV. With $2n$ odd nodes the path may be described by n distinct routes.

The Königsberg diagram has 4 odd nodes and thus requires two separate routes to be covered, such as *CABACDA* and *BD:* diagram 1 has no odd nodes and is unicursal: diagrams 2 and 3 have each just two odd nodes and can be traversed from one to the other.

Check these rules against any drawn network. See how far you can go towards finding proofs for the propositions, starting by considering very simple patterns of lines and then attempting to develop the complexity. You now know *how* to test a particular network, but note that the verification of the facts does not, itself, *prove* anything. You should seek out proofs for yourself if you have sufficient interest.

14. Topology: (b) Möbius

TOPOLOGY deals with the essential qualities of shapes: it is a quintessential Geometry. Just as the Quintessence was the fundamental essence, virtue, power or material within the Elements of Earth, Air, Fire and Water, so Topology is the warp and weft upon which any tapestry of Geometry is woven.

A network is made up of connections between the nodes; it is not concerned with the nature of the branches making the connections. Topology deals with the *connectivity* of points, not with the details of the connection. Any deformation of a shape, such as by stretching or bending, will not alter the topological nature of the shape so long as the connections continue to exist between points of the domain of the shape. If a connection exists, a cut will break it and separate the ends into two domains. In (i) a single cut will separate domains A and B; in (ii) a double cut is necessary and (iii) needs three cuts to separate A and B.

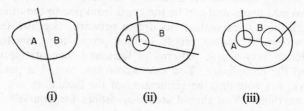

| (i) | (ii) | (iii) |

LONG STRIPS OF PAPER, SCISSORS AND SELOTAPE ARE NEEDED HERE

Take a long strip of paper and fold it into a ring or cylinder, sticking the ends together. There are clearly two domains of points, two surfaces, one on each 'side' of the paper, which are unconnected. If you were a 2-dimensional ant living on one side you could not know of a similar ant living on the other side: a 3-dimensional sense would be needed to enable you to crawl around the edge.

Now take a long strip of paper and give it *one half twist* before sticking the ends together. Mark a point X on one 'side' of the paper and a point *Y* behind it. Imagine, again, you are the 2-dimensional ant living at *X* and another lives at *Y*. Can you meet? You will find you can.

The zero-twist world, then, consists of *two* surfaces with *two* edges, but the one-half-twist world is but *one* surface with *one* edge. Cut down the middle of the first world: clearly the two existing domains become duplicated. Cut down the middle of the second world. Connections have been cut. In what way are there now two domains? Answer this last question with care!

Next, attempt to cut the one-half-twist into three parts. The join shown should be cut along *Aa* with the aim, later, of cutting along *Bb*. Record carefully what happens, and attempt to explain it.

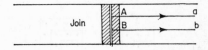

Take a further long strip and stick the ends together after *two* half twists. Make the *X*, *Y* test for number of surfaces: cut into two and again note the numbers of surfaces and edges and the nature of the cutting of connections.

The intriguing results are here for *you* to find: see what conclusions you can come to about these surfaces, called Möbius strips after the German Mathematician of the nineteenth century, A. F. Möbius.

Your results might, perhaps, be tabulated:

half-twists	cuts	strips	surfaces	edges
0	0	1	2	2
0	1	2	4	4
1	0	1	1	1
1	1			
1	2			
2	0			
2	1			

The single half-twist surface has one edge only. A dress designer might try stitching a sleeve of this type to an armhole. Unfortunately, however, this can only be done in *four* dimensions!

15. Some Properties of Numbers

PROPERTIES of Numbers! The title is too grandiose for this galli-maufray or hodge-podge of facts! Each of the nine sections below allows room for extension by personal investigation: many of the facts will probably be already known.

1. The sum of the first n odd numbers is n^2. This is easily shown diagrammatically:

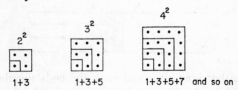

Try to imagine a 3-dimensional form of these squares. Find the missing terms in the following sum, and give the value of the sum.

$$1 + 7 + 19 + 37 + \ldots + 271$$

2. The sum of the first n natural numbers is $\frac{1}{2}n(n + 1)$.

Let $\qquad S_1 = 1 + 2 + 3 + \ldots + (n - 1) + n$

then $\qquad S_1 = n + (n - 1) + (n - 2) + \ldots + 2 + 1$

By addition,

$$2S_1 = (n + 1) + (n + 1) + (n + 1) + \ldots + (n + 1) + (n + 1)$$
$$= n(n + 1) \quad \text{and the result follows.}$$

If S_2 is the sum of the squares of the first n natural numbers and S_3 the sum of their cubes, the formulae are:

$$S_2 = \frac{n}{6}(n + 1)(2n + 1) \quad \text{and} \quad S_3 = \frac{n^2}{4}(n + 1)^2$$

But the proofs are rather more involved than that given for S_1. Verify these in particular instances and note the interesting fact

$$S_3 = S_1^2$$

Consider as a case of this, that using the S_3 formula

$$1 + 8 + 27 + 64 + 125 = \frac{5^2}{4}(5 + 1)^2 = \frac{30^2}{4} = 15^2$$

But the series can be written

$$1 + (3 + 5) + (7 + 9 + 11) + (13 + 15 + 17 + 19)$$
$$+ (21 + 23 + 25 + 27 + 29)$$

which is the sum of the first 15 odd numbers and equal to 15^2 as in section one above. The sum of the first five natural numbers is 15, as shown earlier.

3. Sum of the reciprocals of the natural numbers: the Harmonic series. There is no formula for

$$1 + \frac{1}{2} + \frac{1}{3} + \frac{1}{4} + \ldots + \frac{1}{n}$$

The sums for the first few terms are 1, $1\frac{1}{2}$, $1\frac{5}{6}$, $2\frac{1}{12}$, $2\frac{17}{60}$ . . . but there is no *limit* to the answer. By counting far enough the sum can exceed *any* stated number. It can exceed 5 or 10 or 2 million! Thus both

$$1 + 2 + 3 + 4 + \ldots \quad \text{and} \quad 1 + \tfrac{1}{2} + \tfrac{1}{3} + \tfrac{1}{4} + \ldots$$

reach to any total desired provided sufficient terms are taken.

The series $1 + \tfrac{1}{2} + \tfrac{1}{4} + \tfrac{1}{8} + \ldots$ however, whose terms get successively smaller and smaller as also do those of the previous series, does have a *limit*. The sums are 1, $1\frac{1}{2}$, $1\frac{3}{4}$, $1\frac{7}{8}$, . . . and can be inferred to continue as $1\frac{15}{16}$, $1\frac{31}{32}$, . . . and generally as $2 - \tfrac{1}{2}^n$ after $n - 1$ terms have been counted, tending to a *limit* of 2. The flea which jumps 1 ft. then $\frac{1}{2}$ ft. then $\frac{1}{4}$ ft., halving each time, can never reach his destination 2 ft. from his starting point!

4. Pyramidal numbers.

The numbers S_1 can be represented as triangles.

| 1 | 3 | 6 | 10 | $\frac{1}{2}n(n + 1)$ |

The number of cannon balls in a triangular pyramid pile is the sum of these numbers, 1 ball on top, 3 in the second layer, 6 in the next, until, at the base there were first put down some number of the form of $\frac{1}{2}n(n + 1)$ balls. The total can be shown to be $\frac{1}{6}n (n + 1)(n + 2)$. These are the *triangular pyramidal* numbers 1, 4, 10, 20, 35, 56, . . .

The analagous *square pyramidal* numbers are the sums of

| 1 | 4 | 9 | 16 | n^2 |

These are S_2, already given in section two as $\frac{1}{6}n (n + 1)(2n + 1)$ namely, 1, 5, 14, 30, 55, 91, . . .

There is one number *only* which is both square and square pyramidal. You need 4900 grapefruit if you want to display them either in a square pyramid or in a square frame!

5. Primes.

Integers, or whole numbers, are *composite* if they can be built up as the product of smaller integers, or *prime*, if not. 1 does not 'build' other numbers, so is not considered in either class. The first few primes are therefore 2, 3, 5, 7, 11, 13, 17, 19, 23, . . . There is no highest prime, for, if P is the product of all primes less than a stated n, then $P + 1$ must be prime or a multiple of a prime greater than n. In either case there is a further prime. There is no general way of finding where the next prime is: the Sieve of Eratosthenes—the throwing out from the set of natural numbers of the multiples of 2, then 3, then 5, 7, 11, . . . is only really practicable for the smaller primes. The highest I have read of, is $2^{4423} - 1$, which contains 1332 digits!

There is no formula for the number of primes less than n, but if this is denoted as $p(n)$ then there is an approximation,

$$p(n) \doteqdot \frac{n}{\log n}$$

The logarithm here is a 'Natural' logarithm, not the base 10 log used for calculation. Try the formula for a known result, such as taking n as 100, making a list of the primes from 2 to 97—note $91 = 7.13$—and comparing the number of primes so obtained with $p(100)$.

$$p(100) \doteqdot \frac{100}{\log 100} = \frac{100}{4 \cdot 605} = 21 \cdot 72$$

(Some sets of 4-figure tables will be found to contain Natural logs, or they may be obtained by dividing the 'ordinary' log by ·4343.) You might also use, as a test case, the fact that there are 664580 primes less than 10 million.

6. Perfect Numbers.

A number is *perfect* if it is the sum of its own factors, including 1. They were supposed to have mystic significance. The first four are 6, 28, 496 and 8128, since $6 = 1 + 2 + 3$ and $28 = 1 + 2 + 4 + 7 + 14 \ldots$ Verify the others!

7. Divisibility.

The number *written* as $ab \ldots klm$ is divisible by the highest digit of the Scale of Notation if $a + b + \ldots + k + l + m$ is divisible by that digit.

The value of the written number, say N, in scale r is

$$N = m + lr + kr^2 + \ldots + ar^n$$

Let us write, where Q is quotient and R remainder,

$$\frac{N}{r - 1} \equiv Q + \frac{R}{r - 1}$$

Then
$$N \equiv Q(r - 1) + R$$

This is an *identity* true for *any* value of r. Taking r as 1, we have

$$R = m + l + k + \ldots + b + a$$

It follows that if N is exactly divisible by $(r - 1)$ that either the sum of the digits is zero, which is trivial, or *any multiple* of $(r - 1)$.

Thus there is a simple test for dividing by 9 in our Denary scale or for 4 in the Quinary scale, for instance. Writing $M(n)$ to mean a *multiple* of n, we have:

Denary scale $1 + 5 + 8 + 4 = M(9) \Rightarrow 1584 = M(9)$
Octal scale $2 + 2 + 2 + 1 = M(7) \Rightarrow 2221 = M(7)$
Scale three $1 + 2 + 2 + 1 = M(2) \Rightarrow 1221 = M(2)$
Duodenary scale $9 + E + 0 + T \neq M(E) \Rightarrow 9EOT \neq M(E)$.

Verify these and others of your choosing. A particular instance such as the first above may be proved this way:

$$1 \cdot 10^3 + 5 \cdot 10^2 + 8 \cdot 10 + 4$$
$$= 1 \cdot (999 + 1) + 5(99 + 1) + 8(9 + 1) + 4$$
$$= M(9) + (1 + 5 + 8 + 4)$$

which is a multiple of 9 since the second bracket is a multiple of 9. There is an easy proof using *congruence:* try to find out about this.

8. Pythagorean integers.

The triad (3, 4, 5) consists of the three smallest integers which compose a right plane triangle. Put another way, (3, 4, 5) is the

lowest integral solution of the Pythagorean condition for a plane right-angled triangle,

$$x^2 + y^2 = z^2$$

Let $x = p^2 - q^2$ and $y = 2pq$. Work out $x^2 + y^2$. You find this to be $p^4 + 2p^2q^2 + q^4$. Now this is the square of $p^2 + q^2$. So with this x and y, we have $z = p^2 + q^2$. Hence the triad ($p^2 - q^2$, $2pq$, $p^2 + q^2$) *always* satisfies the equation. These values will always provide integral right triangles if p and q are themselves integral.

p	q	Sides of triangle
2	1	(3, 4, 5)
3	2	(5, 12, 13)
4	1	(15, 8, 17)

and others as you care to determine, without limit.

The solution of *indeterminate* equations such as this belongs to what has been called Diophantine analysis after Diophantus, an Alexandrian of the third century A.D. Pierre Fermat, a 17th century Scholar and Councillor of Toulouse and a great amateur Arithmetician, published many discoveries without proof. In his copy of Diophantus' Algebra he wrote in a margin that the equation

$$x^n + y^n = z^n$$

has *no* solution in integers except for $n = 1$ or 2. This 'Last Theorem' or supposition of Fermat has never been proved nor disproved!

9. Patterns.

$$0 . 9 + 1 = \quad 1$$
$$01 . 9 + 2 = \quad 11$$
$$012 . 9 + 3 = \quad 111$$
$$0123 . 9 + 4 = 1111$$

Continue this. Can you justify it? (Try multiplying $1 . 10^2 + 2 . 10 + 3$ by $10 - 1$). Is the pattern accidental or representative of structure? Can you find some other such patterns?

16. Circulating Decimals: (a) Properties

YOU WILL, at some time or other, have met in decimal division a repeating pattern of digits after the decimal point such as 26·333 . . . or 2·8555 . . . or ·507507507 . . . but suppose you have a decimal like this in front of you but not the division which gave rise to it. How could you find it? Doubtless you know that the first example above is $26\frac{1}{3}$. For the next suppose we write $y = \cdot 8555$ . . . and by multiplying by 10 separate the decimally repeating part. $10y = 8 + \cdot 555$ . . . The recurring decimal can then be dealt with thus:

If $x = \cdot 555$ . . .
then $10x = 5 \cdot 555 . . . = 5 + x$
so $9x = 5$
and $x = \frac{5}{9} \Rightarrow 10y = 8\frac{5}{9}, y = \frac{77}{90}$

Finally we see $2 \cdot 8555 . . . = 2\frac{77}{90}$

Let us look at the last example above and write this time
$$x = \cdot 507507507 . . .$$
clearly $1000x = 507 \cdot 507507 . . . = 507 + x$
or $999x = 507$
whence $x = \dfrac{507}{999} = \dfrac{169}{333}$

It is unlikely you ever worked out this division! Do it now by long division to convince yourself the fraction is correct.

$$
\begin{array}{r}
\cdot 5 \\
333 \overline{)169000000} \\
1665 \\
\hline \\
250 \quad \text{and so on.}
\end{array}
$$

Now choose some other repeated patterns of digits and find the division, or the fraction, which led to those patterns. You will soon see that a pattern $\cdot abc \ldots kabc \ldots kabc \ldots k \ldots$ derives from the fraction $\dfrac{abc \ldots k}{999 \ldots 9}$ which may not be in its simplest form. Thus

$\cdot 333 \ldots = \frac{3}{9} = \frac{1}{3}$ and $1\cdot 0606 \ldots = 1\frac{06}{99} = 1\frac{2}{33}$ and $\cdot \dot{0}243\dot{9} = \frac{2439}{99999} = \frac{271}{11111}$ and this is $\frac{1}{41}$!

We are going to look at some of these decimals deriving from fractions with a prime denominator so we might consider here $\dfrac{4}{13}$ or $\dfrac{27}{73}$ but not specifically $\dfrac{8}{21}$ nor $\dfrac{101}{36}$.

Write out $\dfrac{1}{7}$ as a decimal. You can write $\dfrac{1}{7}$ as $\dfrac{1\cdot 0000000}{7}$ and divide until the shape of the result becomes clear.

$$\tfrac{1}{7} = \cdot 14 \ldots \text{ and so on.}$$

You will have found the digits 142857 repeating themselves over and over again, the '1' returning as soon as the remainder is 1 so that 10 has to be divided by 7 again as at the beginning of the division. Now 999999 is a multiple of 7 so 1000000 when divided by 7 has a remainder of 1 and the repetition starts then.

$$\frac{999999}{7} = 142857 \qquad \frac{1000000}{7} = 142857 + 1$$

$$\text{and } \frac{1}{7} = \cdot 142857 \text{ rem. } 1$$

(You should try to tie up in your mind this sequence of nines with the denominator sequence above.)

We say the decimal $\cdot 142857142857 \ldots$ is *periodic* and *circulates*.

Its period is 6 as six digits circulate and its cycle is 142857. We can represent the cycle thus: The upper 1 shows the start of the cycle and the arrow shows the direction.

For simplicity, refer to $\frac{1}{7}$ as 142857, it being understood that this is the cycle of the decimal expression of this fraction. Now find the cycle for all possible sevenths, either by division or more easily by multiplying 142857 by 2, 3, 4, 5, 6. (Since 142857 × 7 < 10⁶ there can be no danger in this method; each cycle obtained by multiplication will remain as six digits.) Put your results into a Table where the side numbers are the numerators of the sevenths.

7th						
1	1	4	2	8	5	7
3	4	2	8	5	7	1
2	2	8	5	7	1	4
6	8	5	7	1	4	2
4	5	7	1	4	2	8
5	7	1	4	2	8	5

Note the characteristics of this Table carefully. The side numbers—which you notice may be written *along the top*—are the successive remainders: this perhaps makes it clear why the pattern is *cyclic*. Turning $\frac{1}{7}$ into a decimal involves dividing 10 by 7, leaving a remainder of 3 leading to $\frac{3}{7}$, requiring division of 30 by 7, leaving a remainder of 2 leading to $\frac{2}{7}$, needing 20 to be divided by 7 with a remainder of 6 . . . and so on.

To express any seventh as a decimal, *only* the first digit need be calculated, the rest follows. You do not even need to know all the cycle, only half of it! Diametric digits are *complementary*—adding up to 9.

From what you did earlier you would not expect 7 to be unique in its behaviour. It is merely the first Prime to show these characteristics clearly. 2 and 5 are both factors of ten and therefore terminate as decimals and 3 has 2 periods of 1 digit each ($\frac{1}{3}$ = ·3, $\frac{2}{3}$ = ·6) but this is not very exciting. Go on then to elevenths. Write out the cycle for $\frac{1}{11}$, $\frac{2}{11}$, . . . as 09, 18, 27, . . . until you have all ten, and attempt cycle diagrams and tables as before. There are only 2 digits to circulate—a rather grandiose term here! We have

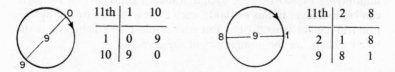

11th	1	10
1	0	9
10	9	0

11th	2	8
2	1	8
9	8	1

and *three more*. Complete these cycles and tables. You have found that 11 has 5 cycles with a period of 2 digits each.

Collecting facts so far:

Prime	Number of Cycles	Period	Product of col. 2 and 3
3	2	1	2
7	1	6	6
11	5	2	10
13			
17			

we can guess the possible entries for 13 and 17, the actual period depending, you remember, on the number of nines, 999 . . . 9 needed before a multiple is found. Now complete the cycle and table diagrams for 13.

17. Circulating Decimals: (b) Calculation

YOUR INVESTIGATION in the last chapter suggests that all divisions by a prime number, which do not finish at some stage of the work, produce a pattern of digits which endlessly circulate. The *period* of the repetition is a factor of the *prime less one*. The product of the period and the number of different *cycles* is the *prime less one*. Also, if the period is even, the first half of the cycle is complementary to the second half. We have four results:

A. Quotients terminate or circulate.
B. When the divisor is the prime p, the period is a factor of $p - 1$, e.g. 7 leads to a period of 6, and 37 to one of 3.
C. The product of the period and the number of different cycles is $p - 1$, e.g. 11 gives a period of 2 but there are 5 different cycles. The product of 2 and 5 is 10, one less than 11. A cycle exists, therefore, for each of the 10 fractions $\frac{1}{11}$ to $\frac{10}{11}$.
D. If the period is *even* the first half of the cycle is complementary to the second half, e.g. $\frac{2}{13} = \cdot 153\ 846\ .\ .\ .$

These rules shorten the labour needed to compute a repeating division. You must work by hand or use a calculating machine. If you use the latter, practise continuing the division beyond the limit imposed by the size of the machine's register, by noting the *remainder* at the furthest stage and then resetting to divide this remainder by the chosen prime. Thus dividing 1 by 17 by successive subtraction, the calculator shows 058823 with a remainder of 9. Resetting to divide 9 by 17 then gives 529411 remainder 13, so

$$\tfrac{1}{17} = \cdot 058823529411\ .\ .\ .$$

By B the period must be 2, 4, 8 or 16 and clearly can now be seen to have to be 16. By D the whole cycle may be written down *at once* following the 8th digit:

0588 2352 9411 7647

Use this help with awareness and caution: the period must be even and a half-period completed before the complement can be written down. You will find $\frac{1}{41} = \cdot 02439 \ldots$ but the 9 is *not* the start of the second half! In fact

$$\frac{1}{41} = \cdot 02439\ 02439\ 02439 \ldots$$

There is an *odd* period of 5 digits, 02439 and seven other cycles 04878, 07317, 09756, 12195 . . . and so on.

Here is a table for the first few primes. Verify some of the results for yourself. Find the cycles and prepare the diagrams and squares, as in the last chapter, in suitable cases.

Prime	Period	No. of cycles
3	1	2
7	6	1
11	2	5
13	6	2
17	16	1
19	18	1
23	22	1
29	28	1
31	15	2
37	3	12
41	5	8
43	21	2
47	46	1
53	13	4
59	58	1

You know the cycle $abc \ldots k$ derives from the fraction

$$\frac{abc \ldots k}{999 \ldots 9}$$

Factors of 9, 99, 999, . . . will thus signify the primes leading to the smaller periods 1, 2, 3, . . .

No. of 9's	Denominator	Prime divisor	Period
1	9	3	1
2	99	11	2
3	999	37	3
4	9999	101	4
5	99999	41 and 271	5
6	999999	7 and 13	6
7	9999999	239 and 4649	7

Can you factorise higher sequences of 9's? Show that 73 is a factor of 99,999,999. What other prime division leads to a cycle of 8 digits? The only factors of 99,999,999,999,999,999,999,999 are 9 and 11,111,111,111,111,111,111,111 so this prime consisting entirely of the digit 1 entails a period of 23 by division! Incidentally, it shows that 11,111,111,111,111,111,111,110 is a multiple of 23! Verify this!

You may extend your investigation to division by composite numbers, if you wish, and discover and analyse such results as

$$\tfrac{1}{49} = \cdot 02\ 04\ 08\ 16\ 32\ 64\ .\ .\ .$$

18. Magic Squares

A MAGIC SQUARE is an array of whole numbers, often 1 to n^2, arranged in the form of a square so that the sum of every row, column and diagonal is the same.

8	1	6
3	5	7
4	9	2

Here the numbers 1 to 9 are placed so that the constant sum is 15. This particular magic square was used as a charm and was known in China long before 2000 BC. It has a side of *three* numbers and is of the *third* order. Clearly, other third order squares can be found from this one by the addition of the same number to each element. The subtraction of 1 from each element leads to a square with a line-sum of 12 containing the numbers 0 to 8.

7	0	5
2	4	6
3	8	1

Now, these figures 0 to 8 can all be represented in Scale Three by *two* digits, namely 00, 01, 02, 10, 11, 12, 20, 21, 22.

21	00	12
02	11	20
10	22	01

The reason for the 'magic' quality is much clearer here. Each row and column consists of a 0, 1 and 2 in both the first and second digit position. The square follows the superimposition of two simpler arrays. One diagonal likewise consists of 0, 1 and 2, but the other contains the same total with 1, 1 and 1.

2	0	1
0	1	2
1	2	0

1	0	2
2	1	0
0	2	1

A French Jesuit, Simon de la Loubère, was envoy to Siam—Thailand—towards the end of the seventeenth century. On his return to France he wrote extensively about the Kingdom of Siam. His work was translated into English in 1693. In the book he reported a method for the construction of magic squares of *odd* order. The rules for construction are, in effect:

A. Put 1 in the middle of the upper row.

B. Proceed *cyclically* ⟋ .

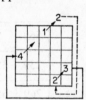

 C. If A and B cannot be followed, go to the position below the last placed number and place *there* the next and then continue as before.

Complete the square of order five and prepare that of order seven. Check the totals of the rows, columns and diagonals. The total is easily calculated: the sum of the integers 1 to n^2 is $\frac{1}{2}n^2(n^2 + 1)$, but this is the sum of n rows or columns. The line-sum is thus $\frac{1}{2}n(n^2 + 1)$ for a magic square of order n.

Mistakes in your work are unlikely to be soon discovered, especially if you attempt squares of higher order—so take care! The final number n^2 will appear in the middle of the bottom row if the construction has been correct.

Reduce each of the elements of the order five square by one, and rewrite it in the Quinary scale with the numbers 00 to 44. Do the same with the seven square using Scale Seven and 00 to 66. Compare the results with those of the Chinese square.

The distinguished German artist Albrecht Dürer incorporated a magic square of order four into his engraving 'Melancholia' done in 1514. The square shows the date!

16	3	2	13
5	10	11	8
9	6	7	12
4	15	14	1

More recently, in 1961, a magic square of order *thirteen* was formed consisting entirely of *prime* numbers! A rather smaller such square is shown here.

569	59	449
239	359	479
269	659	149

de la Loubère's construction is, of course, but one way of forming just one type of Magic Square.

19. Finite Arithmetic

ARITHMETIC means, literally, the art or technique of counting. To a child it may be all of mathematics, to many a man an ability to check a bill quickly and correctly. Arithmetic deals with the properties of numbers and their manipulation under the operations of adding and multiplying and their *inverse* operations, subtraction and division. The problem, 'How many pencils at 7d. each can be bought for 3s. 6d.?', is essentially the question, 'What is the number x such that $7x = 42$?'. The answer is 6. But what of the similar question, 'How many pencils at 7d. each can be bought for 3s. 9d.?' I suppose the answer is still 6. If you frame the question, remembering that fractions of a pencil cannot be bought, as, 'What is the integer x such that $7x = 45$?', the only response is, 'There is no answer'. Such problems can typify the nature of an Arithmetic. Can a problem be solved with the numbers available to provide solutions? Is there an x such that $x + 3 = 2$? It all depends what numbers are available.

The 'ordinary' arithmetic deals with an unlimited set of numbers, whether or not there are included, fractions, negatives, irrationals . . . and so on. The number set is in-finite. Here we are looking at Finite Arithmetic where the number set consists of a specific *finite* set of numbers and those only. Thus we shall be considering the behaviour of the set $\{0, 1, 2, 3, 4\}$ and the solubility of problems under addition, and of the set $\{1, 2, 3, 4, 5\}$ and the solubility of problems under multiplication.

Finite number sets arise when only *remainders* are used in 'ordinary' operations. Thus, to take four examples:

(i) $2 + 4 = 6$ but if each number is *reduced modulo* 5, that is, written as its remainder after division by 5, then $2 + 4 = 1$.

(ii) $12 - 15 = -3$ but reduced modulo $\underline{4}$ we have $0 - 3 = 1$. (Note $-3 = -4 + 1$).

(iii) $3 . 7 = 21$ but $1 . 1 = 1$ (mod 2).

(iv) $\dfrac{26}{13} = 2$ but $\dfrac{2}{1} = 2$ (mod 6).

77

The easiest approach to such arithmetic—Finite Modular Arithmetic, to give it its full title—is to imagine the set depicted on a clock face. In everyday usage $10 + 4 = 2$ on a clock: this is Finite Arithmetic mod 12.

For the first set exemplified above, we have:

and the addition table

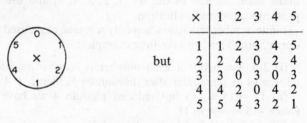

+	0	1	2	3	4
0	0	1	2	3	4
1	1	2	3	4	0
2	2	3	4	0	1
3	3	4	0	1	2
4	4	0	1	2	3

Any number chosen is picked from the *top* and the set number to be added, from the *side*. The result may be had by adding around the 'clock'. Is there any addition that cannot be performed with the numbers? Is there any problem that cannot be solved? Given *any* two numbers a, b of the set, the same or different, is it always possible to find *uniquely* a member x of the set such that $a + x = b$? You will quickly see this is so. We say

$$\forall\, a, b \in P\, \{0, 1, 2, 3, 4\}\, \exists\, x \in P : a + x = b$$

or, 'for all a and b which are members of the set P containing the numbers 0, 1, 2, 3, 4 there exists a number x, also a member of P, such that the sum of a and x is b'.

Try now the clock face with 0 to 5 upon it, and the operation of multiplication. There is little to be gained by writing the zero products of zero. We then have

but

×	1	2	3	4	5
1	1	2	3	4	5
2	2	4	0	2	4
3	3	0	3	0	3
4	4	2	0	4	2
5	5	4	3	2	1

Check each of the entries of the table, e.g. for 5 . 2, count 5 around the clock from 1 to 5, and then 5 again, finishing at 4. Can multiplication be performed between all numbers of the set? Are problems soluble? Here are some examples of the set's behaviour under multiplication.

(i) Does $3 . 4 = 4 . 3$? Yes. Both are zero.

(ii) Does $1 . 2 . 4$ have an unique meaning? Yes. The result gained by any order of working is always 2.

(iii) What is the value of $\frac{3}{4}$? This is asking, 'Is there an x such that $4x = 3$?'. The table shows there is *no* solution. $\frac{3}{4}$ has *no* meaning in the arithmetic being dealt with.

(iv) What is 4 divided by 2? The table assures us that $2 . 2 = 4$ *and* that $2 . 5 = 4$. The answer then is 2 *or* 5!

(v) If a is a member of the set, then $a . 0 = 0$. Can $a . b = 0$ when $b \neq 0$? Yes. $2 . 3 = 3 . 2 = 0$ and $3 . 4 = 4 . 3 = 0$.

(vi) What is the value of 3^n where n is a member of the set? Answer: always 3. (What about 3^0?)

(vii) Which members of the set have square roots? This is asking, 'For what a is there a solution of $x . x = a$?' We see, there are, for $a = 1, 4, 3$, the numbers in the *dexter* diagonal.

(viii) Solve $4x = 0$. Solution $x = 0$ or 3.

(ix) Solve $(x - 1)(x + 2) = 0$. Each factor may be zero, so that $x = 1$ or -2, i.e. 4: one factor may be 2 if the other is 3, *or* one 3 if the other 4. Neither of these is possible, so the only solutions of $x^2 + x - 2 = 0$ are $x = 1$ or 4.

(x) Solve $x(x - 1)(x - 2) = 0$. Clearly 0, 1, 2 are roots. $x = 3$ leads to $3 . 2 . 1 = 0$ and $x = 4$ leads to $4 . 3 . 2 = 0$ and $x = 5$ to $5 . 4 . 3 = 0$. Thus *every* member of the set is a root of this cubic equation $x^3 - 3x^2 + 2x = 0$!

Make up further examples and convince yourself that this type of arithmetic is both Commutative and Associative under addition and multiplication, but that problems may have *no* solution, or an *unique* solution or *many* solutions according to the details of the problem.

In mod 5 the multiplication table is:

×	1	2	3	4
1	1	2	3	4
2	2	4	1	3
3	3	1	4	2
4	4	3	2	1

Here, convince yourself *every* problem is soluble. There is, surely, therefore, a fundamental difference between the mathematical *structure* of mod 5 and mod 6 arithmetic under multiplication.

Try problems involving both tables in these and other moduli, such as 3, 4, 7 or 10. Verify that the addition table always gives an unique solution to problems but that the multiplication table only does so for a *prime* modulus. Does the Distributive law hold in these arithmetics—for multiplication over addition, and for addition over multiplication?

An Arithmetic where addition, subtraction, multiplication and division is always possible—apart from division by zero—is called a *field*. An Arithmetic with addition, subtraction and multiplication always possible, but not necessarily division, is a *ring*. Thus, for instance, we have seen arithmetic modulo 5 is a field and arithmetic modulo 6 a ring. Is 'ordinary' arithmetic either of these?

20. Latin Squares: Idea of a Group

A NEAT-MINDED child has a box of sixteen coloured blocks. There are four of each of the colours, red, green, yellow and blue, and the box contains them four by four. The child, being tidy and precise, puts them away so that one block of each colour is in each row or 'column' of the box. His pattern of blocks happens to be as shown. Such a pattern is a *Latin square*. How many such arrangements are there, of *order* four like this? The answer is 4 . 4! . 3! which is 576—a surprisingly high number, you may think. Try to find the number of arrangements of order two or order three.

For our purpose, let us consider the number of *essentially different* patterns when the top row and left column are the same, as in the example illustrated. The *pattern*, as opposed to the arrangement, is not different if colours are interchanged, yellow for blue, say, so that there are 4! arrangements of any one pattern. How many different patterns are there? Suppose the colours be represented by *A*, *B*, *C*, *D*. How many different patterns are there, with *ABCD* in the top row, *ABCD* in the left column, and one of *A*, *B*, *C*, *D* in each row and column? Experiment with this and I expect you will decide there are four possibilities, I, II, III and IV. The earlier example is type IV.

I	II	III	IV
A B C D	A B C D	A B C D	A B C D
B C D A	B D A C	B A D C	B A D C
C D A B	C A D B	C D B A	C D A B
D A B C	D C B A	D C A B	D C B A

Are these four truly distinct? Try interchanging C and D in I or B and D in II, and then re-arranging the *complete* rows or columns to return to the ABCD frame of the top row and left column. Clearly the labelling is arbitrary: the sets {red, green, yellow, blue}, {A, B, C, D}, {1, 2, 3, 4}, {*p, a, t, z*} or *any* set of four elements can label the positions within the square. Patterns I, II, and III all have the same essential shape; they are *iso-morphic*. What interchange converts III to I?

We are left finally with just *two* patterns, the *Cyclic* (for obvious reasons), and the *Klein*.

A B C D	A B C D
B C D A	B A D C
C D A B	C D A B
D A B C	D C B A
Cyclic	Klein

The investigation of such arrangements and patterns is itself a nice piece of Mathematics, but its relevance is much greater than is obvious. With the set {1, 2, 3, 4} *and the operation of multiplication*, where have you met the pattern Cyclic II?

Now consider the set $\{1, i, -1, -i\}$ where $i^2 = -1$ and the operation of multiplication. Complete the table:

×	1	i	−1	−i
1	1	i	−1	−i
i	i	−1	−i	
−1				
−i				

Where else, in the last chapter, did you encounter this Cyclic I pattern?

Next, take a postcard and make a hole in the top right corner. Draw a rectangle to frame the card. Take the card from the frame and then put it back in any way desired, the hole now being at

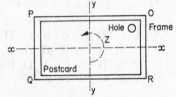

O, P, Q or R. The single operations leading to these positions we shall label:

I. The card is returned to the position it was in.

X. The card is half turned about the X axis.

Y. The card is half turned about the Y axis.

Z. The card is half turned about the centre, in its own plane.

Suppose one such operation followed by another is shown as $X° Y$, this meaning, operation X *is followed by* operation Y. Perform the sequence of operations for yourself and discover that the two operations have the same result as the single operation Z. $X° Y = Z$. Does $Y° X = Z$? Is the *Group* of *these* operations Commutative? Verify the table of results, the top line operation being followed by the left column operation. It is pattern IV, the second distinct pattern, the Klein.

o	I	X	Y	Z
I	I	X	Y	Z
X	X	I	Z	Y
Y	Y	Z	I	X
Z	Z	Y	X	I

The behaviour, then, of *certain* systems can be displayed in a particularly neat table. The characteristics displayed are that with a set of elements and an operation, say, *:

1. The result of operating on two members of the set—a *binary* operation—is to reach another member of the set. $a*b = c$.
2. The Associative law holds. $a*(b*c) = (a*b)*c$.
3. There is an *unity* element such that the operation produces no change. $a*I = I*a = a$ for every member of the set.
4. There is an *inverse* element corresponding to each element. For each '*a*' there is an element a^{-1} such that $a*a^{-1} = a^{-1}*a = I$.

This seems terribly technical, doubtless! We have, so far, seen various systems which are representative of what are called *Group* structures:

> All modular arithmetic under addition. (Cyclic I.)
> $\{1, i, -1, -i\}$ under multiplication. (Cyclic I.)
> A *prime* modular arithmetic under multiplication. (Cyclic II.)
> Rectangle positions under rotation. (Klein.)

Let us look at the Group characteristics in these cases.

1. The answer given by the table is in all cases a member of the original set.
2. Consider for yourself the Associative rule in particular instances. As *Groups* are not necessarily Commutative you must work *in the order* of the written symbols. The Associative law provides that *brackets* are not needed, but the *order* of symbols must be kept to.
3. Examples: $2 + 0 = 2$ $1 . i = i$
              $3 . 1 = 3$ $Y . I = Y$

   The unity elements are, respectively, 0, 1, 1, *I*.
4. There is *always* a way of returning to the unity or neutral element. Just as in 'ordinary' arithmetic $8 + (-8) = 0$ or $7 \times 1/7 = 1$ so, with the Groups cited, $4 + (2) = 0$, (mod 6) $i \times (-i) = 1$, $2 . 3 = 1$ (mod 5) and $Y . Y = I$.

These structures are such that every question within the system can be answered; every problem within the system has an unique solution.

The whole-numbers (positive and negative integers and zero), under addition, behave like this, as do the rational numbers under multiplication. Square matrices, such as,

$$\begin{pmatrix} 1 & 0 \\ 0 & 1 \end{pmatrix}, \quad \begin{pmatrix} 0 & -1 \\ 1 & 0 \end{pmatrix}, \quad \begin{pmatrix} -1 & 0 \\ 0 & -1 \end{pmatrix}, \quad \begin{pmatrix} 0 & 1 \\ -1 & 0 \end{pmatrix}$$

behave like this under matrix multiplication: verify this by forming the table. These matrices, in fact, are the set $\{1, i, -1, -i\}$ represented as rotations; it is iso-morphic with the set of rotations $\{0°, 90°, 180°, 270°\}$ under addition. Try to find out the justification for these last remarks!

A Group may consist of any quantity, or of an infinity, of elements having the required characteristics under a certain operation. That all our examples here have been of order four is merely a convenience: that they happen to be Commutative an accident of the choice.

A Group is a set closed under a binary associative operation, possessing a neutral element and an inverse for each element of the set. Suppose the behaviour of a machine, a machine-tool say, can be so represented under its instructions. The Group structure makes it clear the machine will work; will be able to obey instructions. The behaviour of arithmetic modulo 6, under multiplication, should show you what chaos results if the system has *not* Group structure!

21. Loci

THE LOCUS of a point is the set of all the positions the point may have under certain constraints. A weight swung on the end of a piece of string moves in a curve; the string provides the constraint, the curve is the locus.

A locus need not be a smooth curve, a *continuous* line; it may consist of isolated points or of a discontinuous line or series of lines. You have all drawn graphs such as $y = 2x - 1$ or $y = x^2$ where x is measured *across* the page, y *up* the page, and values of y

are found from chosen values of x, the result being shown by the position (x, y). The selected points were joined up to form a smooth line because you realised you could have found more and more points *between* any two already found. You probably did not pause to consider if x could be 2 or -3 or $\frac{3}{4}$ or $\sqrt{2}$ or $\sqrt{-4}$, but such concerns affect the locus. Compare the examples shown with the 'usual' interpretation of $y = x$ and $x^2 + y^2 = 25$.

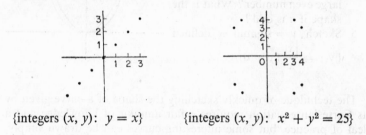

{integers (x, y): $y = x$}   {integers (x, y): $x^2 + y^2 = 25$}

Examples of isolated points

1. If x is a whole number from 1 to 4, plot $y = \sqrt{x}$.
2. Repeat 1 from $x = 1$ to $x = 16$, if x *and* y are whole numbers.
3. Plot $x^2 - y^2 = 0$ if x is a single digit even number.
4. Plot $x^2 + y^2 = 0$.

The symbol $|n|$ means the *numerical* value of n as opposed to the algebraic or signed value. Thus $|3| = 3$ and $|-2| = 2$. The symbol $[n]$ has the value of the highest integer below or equal to n. Thus $[3] = 3$, $[2\frac{1}{2}] = 2$, $[-3] = -3$, $[-2\frac{1}{2}] = -3$. With these symbols the line $y = x$ takes a discontinuous form, although we are now supposing x and y may have any value.

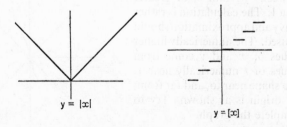

$y = |x|$ $y = [x]$

Examples of discontinuous loci

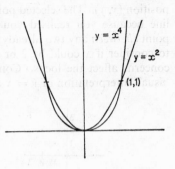

1. Sketch $y = -|x|$.
2. Sketch $y = |[x]|$.
3. Sketch $y = x - [x]$.
4. The shape of $y = x^2$ and $y = x^4$
   is shown. What is the shape of
   $y = x^n$ as n tends to an infinitely
   large even number? What is the
   shape if n is odd?
5. Sketch $y = $ signum x defined
   by $y = \dfrac{|x|}{x}$ $(x \neq 0)$.

The technique of quickly sketching the shape of a curve given by
its Cartesian equation requires a fair amount of knowledge and a
deal of practice, but some interesting curves can be drawn simply,
though with some labour, by plotting points.

Examples of interesting smooth curves

1. The Folium of Descartes. $x^3 + y^3 = 3xy$. Plot points given
   by $x = \dfrac{3t}{1 + t^3}$, $y = \dfrac{3t^2}{1 + t^3}$ for values of t, -2, -3, -10, 0,
   5, 3, 2, 1. For each point (h, k) plotted, plot also the point
   (k, h), the x and y being reversed. Use the same scale for both
   axes, say 2 inches to 1 unit. A suitable range is -2 to 2.

   Draw a *smooth* curve through the points from top left to 0,
   round the loop to 0 again, and down to bottom right.
2. The Swastika, $y^4 - x^4 = xy$.

   Plot points given by $x^2 = \dfrac{t}{t^4 - 1}$, $y = tx$ for various t

   between -1 and 0, or greater
   than 1. The calculation is rather
   heavy and approximation should
   be used. The numerically higher
   values of x and y come from
   values of t numerically near 1.
   The shape near to, and far from,
   the origin is as shown. Try to
   complete the graph.

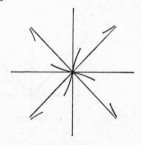

3. The Butterfly, $x^2 = x^6 + y^6$.
   The curve is symmetrical about each of the axes and $|x| \leqslant 1$.
   Choose a very large scale with an x range from -1 to 1 and a
   y range from $-\frac{1}{2}$ to $\frac{1}{2}$.

A point in a plane may also be specified by *Polar* coordinates.
P is the point (r, θ). Draw a 'compass' or *pencil* of lines at angles

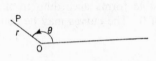
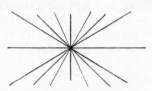

of 30°, 45°, 60°, 90°, 120°, 135° and 150°. Plot selected points on
this, measuring r outwards from 0, and θ anti-clockwise from $0x$,
or the reverse if either is negative.

Examples of Polar Loci

1. Rose petals.   (i) $r = \sin 3\theta$ 3 petals.
              (ii) $r = \sin 4\theta$ 8 petals.

   You need to be able to find the sine of any angle; to know,
   for instance, $\sin 150° = \sin 30°$ and $\sin 300° = -\sin 60°$.
2. The Archimedean Spiral. $r = a\theta$. Choose a constant 'a' to
   suit your scale.

In Polar coordinates 0 is the *pole*.
A system of coordinates with two poles
is *Bi-polar*. The point P is then determined, but not uniquely, by (r_1, r_2).

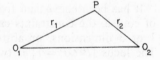

Bi-polar examples

1. Cassini's Ovals. $r_1 r_2 = k$.
2. Cartesian Ovals. $r_1 + cr_2 = k$. This includes the Ellipse
   when c is 1. This oval is easily effected with a taut string
   attached to pins at 0_1 and 0_2.
3. Apollonius' Circle. $r_1 = kr_2$. This is better done at once
   with compasses, if you recall the Angle bisector theorem and
   know where the centre is!

4

In these examples try different constants k and c, and fix chosen
points by the intersection of arcs.

Conchoids and Cycloids

1. The Conchoid of Nicomedes. This curve occurs in the *entasis*
   of an architectural column. XY is a fixed line and 0 a fixed
   point. On any line ON points P and P' are cut off so that
   NP and NP' are a constant length. The locus of P or P' is a
   conchoid. There are three possible forms according to the
   length cut off and the position of 0. The curves may be constructed mechanically.

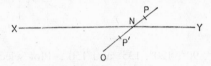

2. Cycloids. If a circle rolls along a line—a wheel along a road—
   the path of a point fixed relative to the circle, is a cycloid. By
   rolling a cardboard circle along various lines, determine the
   shape of various cycloids.
   (i) Along a straight line.
   (ii) Outside a fixed circle.
   (iii) Inside a fixed circle.
   (iv) Inside a fixed circle of diameter *twice* that of the moving
        circle.

It has been convenient to consider a locus in 2 dimensions, but,
of course, the constraints can govern the behaviour of a point in
1 or 3 dimensions, or abstractly, in any number of dimensions.
The locus $\{P : OP = r\}$ is two points in a line through 0, a circle
centre 0 in a plane, a sphere centre 0 or a hyper-sphere centre 0,
according to the type of world in which P may lie.

22. Envelopes

A STANDING FIREWORK throws out its coloured stars in all directions: we shall suppose the explosions send out each particle of flame with the same speed. The scene is a more colourful version of the diagram!

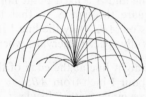

Some sparks fly high and fall near, some go far but rise little, some, projected at about 45°, reach the furthest. All the trajectories seem bounded by the outer surface of the diagram. The outer surface *envelopes* the several curves. The *envelope* of the several parabolas is the outer paraboloid. In a plane section the outer parabola envelopes the others.

The envelope of a wheel rolling along a road is a line parallel to the road: the envelope of a pencil of lines is a point. An envelope may be a point, a line or a surface: it is the configuration determined by the movement of a line, or surface, under certain constraints. The envelope *touches* the moving line in every position that it can take.

A locus in a plane is roughly shown by the position of a number of points which belong to the locus: an envelope is roughly shown

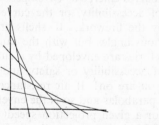

by the position of a number of lines which belong to the envelope. The points and lines above specify the same curve. There is no doubt the envelope gives the clearer picture for the amount of information.

Examples

1. Draw (x, y) axes. Label equidistant points from O as P, Q, R, S, T on one axis and T, S, R, Q, P on the other. Join PP, QQ, etc. The resulting curve is part of a parabola.

2. Draw (x, y) axes and place a ruler across the axes so that a length of 3 inches is cut off between them. Draw, carefully spaced and neatly arranged, six such lines intercepted between the axes, in each of the four quadrants. The attractive envelope is a semi-cubical astroid, a pleasant star shape.

3. AB is a fixed line and S a fixed point not on AB. Place a set square through S so that the right angled vertex is on AB. Only the line 1 is drawn. Do this at carefully spaced intervals, when: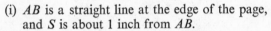

   (i) AB is a straight line at the edge of the page, and S is about 1 inch from AB.
   (ii) AB is a large circle, with S inside.
   (iii) AB is a centrally placed circle of about $1\frac{1}{4}$ inch radius and S is about 1 inch outside.

   The envelopes are a parabola, an ellipse and a hyperbola, but in the last two the drawing should be repeated on the other 'side' of the circle, using a symmetrically placed S.

4. Dogs Alcinous, Balthazar and Calisthenes are at the three corners of a triangular field. At the same moment A starts to chase B, B to chase C and C to chase A. All run at the same speed! What happens? Draw a large triangle and put in successive paths of the chase after short intervals of time. The spiral chase curves emerge as envelopes.

5. The curve of accessibility, or the curve of safety, has been illustrated by the firework. If shells are fired from a fixed point at various angles but with the same muzzle speed, the several trajectories are enveloped by a curve which determines the extent of accessibility or safety under fire—according to which side you are on! If air resistance is ignored, the trajectories are parabolas and so is the envelope. The drawing is not easy. For a given projection speed V, the curve may be plotted as a *locus* in Polar co-ordinates by the equation

$$r = \frac{V^2}{g(1 + \sin \theta)}$$

where g is the acceleration due to gravity. The equation expresses the greatest range on a line inclined at the angle θ to the horizontal through the point of projection.

6. The Cardioid. Draw a circle about $1\frac{1}{4}$ inch radius, near the edge of the paper and mark a point A on it near the edge. With centre at any point P on the circle, draw a circle radius PA. Repeat this many times at carefully spaced intervals all round the first circle. The resulting heart-shaped envelope is the Cardioid.

23. Proof and Paradox

THE WORD *proof* derives from the Latin *probus* meaning *good* or honest. A proof establishes the 'goodness' of a proposition. A paradox is, properly, a proposition contrary to accepted opinion, but the term is often used for an apparently justified self-contradiction. A proof which leads to a paradox is fallacious unless it justifies the paradox. A mathematical proof is often set out line by line: it is a sound proof if the premisses are themselves justified, or are presented as *axioms* or *hypotheses* admitted without proof, and if each line of argument is correctly reasoned from the lines before. A false premiss cannot lead to a valid conclusion however 'good' the argument: a justified premiss is unlikely to lead to a valid conclusion if the argument is faulty; should it, by chance, arrive there, the proof is obviously still worthless; it will not have established the 'goodness' of the conclusion.

The arguments which follow are faulty in premiss or in reasoning: the conclusions are fun—fun is an important part of Mathematics—but the 'proofs' are more than that. Some of them you will probably have seen before; some may amuse you by the unexpected result. The value of such things is to emphasise, through surprise, the importance of mathematical method in reaching a result. The analysis of fallacies can lead to a complexity unwanted here. If you can enjoy the faulty proofs and find wherein the fault lies, in general terms, enough has been achieved! Careful thinking and precise verbal formulation is necessary.

A. $x = y \Rightarrow 2x = 2y$

so half-full = half-empty $\Rightarrow$ full = empty
Thus, a full bottle is empty.

B. Suppose P and Q are two numbers, or any two expressions which can be added, such as arithmetical, algebraical or trigonometrical expressions.

   Prove $P = Q$
   If $P = Q$
   then $Q = P$
   ─────────────────────────────────────
   so $P + Q = Q + P$
   but, in fact, $P + Q = Q + P$
   whence $P = Q$ quod erat demonstrandum

C. Zeno's second paradox.
   The swift running Achilles can never overtake the tortoise in front of him in the race. Before Achilles reaches the point where the tortoise started, the tortoise will have moved on; before he reaches that point the tortoise will have moved on further, and so on for ever: the tortoise can never be reached.

D. Every triangle is isosceles.
   Let the triangle be *ABC. AO* is drawn to bisect angle A and *NO* is drawn to bisect perpendicularly the side *BC. OB* and *OC* are joined and perpendiculars *OX* and *OY* are dropped from O to sides *AB* and *CA*.

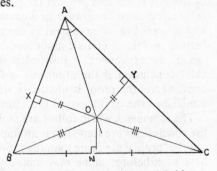

If the triangle is folded along the line *AO*, since *AO* bisects the angle *A*, triangles AOX and AOY are congruent, exactly superimposable,

and $\qquad\qquad\qquad AX = AY \ldots 1$

and $\qquad\qquad\qquad OX = OY \ldots 2$

If the triangle is folded along the line NO, since NO is the perpendicular bisector of BC,

$$OB = OC \ldots 3$$

In the triangles BOX and COY, we have

$$OX = OY \text{ by } 2$$
$$OB = OC \text{ by } 3$$

$\qquad\qquad\qquad$ angles X and Y are each right angles.

Thus the triangles are congruent or equal in all respects,

so that $\qquad\qquad XB = YC \ldots 4$

It follows that $\quad AB = AX + XB$

$$= AY + YC \text{ by 1 and 4}$$
$$= AC.$$

Any triangle is, thus, isosceles.

E. If a positive number is greater than another, then it is also greater than twice that other number.

For, let $\qquad\qquad\qquad a > b$

$$\Rightarrow ab > b^2$$
$$\Rightarrow ab - a^2 > b^2 - a^2$$
$$\text{or } a(b - a) > (b + a)(b - a)$$
$$\Rightarrow a > b + a$$
$$> 2b,$$

which, evidently, proves the proposition.

F. *Solve* $\qquad\qquad 1 + \sqrt{x} = \sqrt{(1 - x)}$

$$\Rightarrow 1 + 2\sqrt{x} + x = 1 - x$$
$$\Rightarrow \sqrt{x} = -x$$
$$\Rightarrow x = x^2$$
$$\Rightarrow 1 = x$$

Check $\qquad\qquad 1 + 1 = 0$

$$\text{or } 2 = 0$$

G. A merchant kept a bag of diamonds as a reserve of wealth. In his will he specified his three sons were to share the diamonds after his death. Ali must receive two-fifths, Peter a third and Benjamin one-fifth. The executor found 56 diamonds of the same weight. 56 is not divisible by 5 or 3, so he borrowed 4 similar diamonds from an accommodating friend and divided the 60 so that Ali received two-fifths, 24, Peter one-third, 20 and Benjamin one-fifth, 12. The 56 diamonds were thus distributed and the 4 borrowed, returned to the accommodating friend!

H. $1 + (1 - 1) + (1 - 1) + (1 - 1) + \ldots$

$= 1 + 1 - (1 - 1) - (1 - 1) - (1 - 1) - \ldots$

$\Rightarrow 1 = 2$

J. Suppose $Z = \dfrac{mx + ny}{x + y}$ where m, n are two different whole numbers.

If $x = 0$, $Z = n$,
If $y = 0$, $Z = m$,
But if $x = y = 0$, then $Z = n = m$,
and since $n = m$ it follows all whole numbers are the same.

K.

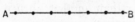   

AB is a line of unit length. It is divided into an even number of equal parts, say $2n$. (The diagram shows the case when n is 3.) Semi-circles are drawn on each successive pair of line segments as shown. Each part of AB has a length of $1/n$ so each semi-circular arc has a length of $\pi/2n$. Since there are n semi-circles, the total wavy length from A to B is $\pi/2n$ times n, or $\pi/2$. The wavy length therefore is *always* the same no matter how many semi-circles are drawn. Suppose the number is high, a million perhaps, then the wavy line AB is very nearly the straight line AB. As the number increases, so ultimately the two become so close that we may say, *in the limit*, the two are the same line.

Thus the wavy length $AB = \dfrac{\pi}{2} = $ the straight length $AB = 1$

$\Rightarrow \pi = 2.$

24. Types of Number

SOLVE the following number equations:

A.	$x - 3 = 0$	B.	$2x - 3 = 0$
C.	$x + 3 = 0$	D.	$2x + 3 = 0$
E.	$x^2 - 1 = 0$	F.	$x^2 + 2x - 3 = 0$
G.	$x^2 - 2 = 0$	H.	$x^2 - 2x - 1 = 0$
J.	$x^2 + 1 = 0$	K.	$x^2 - 2x + 5 = 0$
L.	$x^3 - 1 = 0$	M.	$x^4 - 1 = 0$
N.	$1/x - 1 = 0$	P.	$1/x = 0$

An equation in x is a statement of equality *telling you* something about the number x. The statement $2x - x = x$ is true for all number values of x; it tells nothing about x; it is an *identity*, not an equation. The statement A tells, however, that x has the value 3. We say $x = 3$ is the solution of the equation; that 3 *satisfies* the equation; that 3 is the *root* of the equation.

You may think the selection above a silly one! $A - D$ are so easy, $E - H$ rather a bore, J, K and P just nonsense and L, M and N straight forward enough. But what is meant by solving an equation? No solution to A can be found in the set of even numbers. No solution to D can be found in the set of positive numbers nor in the set of integers. What sets do contain the solutions of $A - D$? What set contains the solutions to all of $A - F$? Give your answers in your own words. The members of the set are *Rational* numbers; numbers which are ratios of whole numbers.

If you knew only the counting numbers, you could solve A, and say the number 1 was a root of E, F, L, M and N; but the other roots of these equations would necessarily be unknown to you. A knowledge of negative numbers is necessary before E and M can be shown to have the root -1, or C and F to have a root -3. Only with a knowledge of fractions can B and D be solved. Equations $E - K$ are all quadratics; are E and F soluble but not $G - K$? If extending the number system, if increasing the membership of the number set, enables A, then B, then C . . . and so on . . ., to be solved, cannot then the others be solved by some further extension? You say $x = \sqrt{2}$ is a root of G and demonstrate with an isosceles set-square to show $\sqrt{2}$. More properly, you add, $x = \pm\sqrt{2}$ and perhaps remember the formula for quadratics and offer $x = 1 \pm \sqrt{2}$ for H. This is already another extension, for $\sqrt{2}$ is not a rational number.

Pythagoras' proof of the irrationality of $\sqrt{2}$ is beautiful and simple. It is necessary to show that the length of the hypotenuse of the set-square is *not* a fractional length as might be expected. It must be a fraction *or* not a fraction. Let us suppose $\sqrt{2}$ *is* the fraction p/q in its lowest terms, so that p and q are whole numbers without a common factor. ($\sqrt{2}$ might be $\frac{7}{5}$, for instance.)

Then	$p^2 = 2q^2$
so	p^2 is even $\Rightarrow p$ is even, equal to $2r$, say.
Then	$4r^2 = 2q^2$
or	$2r^2 = q^2$
so	q^2 is even $\Rightarrow q$ is even.
Whence	p and q have the common factor 2.
But	this is contrary to the supposition
$\Rightarrow$	the supposition is false
	$\sqrt{2} \neq p/q$.

If $\sqrt{2}$ is calculated its value is $1\cdot414213562\ldots$ where there is *no* pattern of circulating digits as with rational fractions. The Greeks were more interested in shape than in number, and thought of number in terms of length: the two come together in the aesthetic concept of a Golden section. If the front of a building is to be a series of columns surmounted by entablature and frieze, what is the best ratio for the height of the columns to the total height? In a Mondrian modern abstract painting, what is the most attractive ratio of the two divided parts of a rectangle?

The Greek answer was that a length is most happily divided if the ratio of the whole to the larger part is also the ratio of the larger part to the smaller. Put a sheet of paper in front of you: where does a crossing line cut the sheet most pleasantly, parallel to a side, in its section of the page into two parts? A $\frac{1}{2}$ cut is clearly too low; $\frac{2}{3}$ seems rather high; $\frac{5}{8}$ seems better. The ancient Greek answer was at the 'end' of the sequence of fractions,

$$\frac{1}{2}, \frac{2}{3}, \frac{3}{5}, \frac{5}{8}, \frac{8}{13} \cdots \frac{a}{b}, \frac{b}{a+b} \cdots$$

Continue the sequence for yourself. If the final ratio is x, then by the definition, x is given by

$$\frac{x+1}{1} = \frac{1}{x}$$

or $x^2 + x - 1 = 0$. If you remember again the formula for a quadratic, you will realise the ratio must be $x = (-1 + \sqrt{5})/2$ or about ·618. The easily written sequence of fractions thus leads to a 'fraction' which is not a fraction at all, but an irrational number! The Greeks used other sequences to approach other irrationals. That for $1/\sqrt{2}$ is

$$\frac{1}{1}, \frac{2}{3}, \frac{5}{7}, \frac{12}{17}, \frac{29}{41}, \cdots \frac{a}{b}, \frac{a+b}{a+(a+b)} \cdots$$

How much more powerful these sequences are, than decimal fractions!

Solutions to equations G and H, then, require the acceptance of irrational numbers. By the same token the solutions of J and K have no meaning unless further interpretation is possible. J gives $x = \pm i$ where $i^2 = -1$, and K leads to $x = 1 \pm 2i$, but have these any meaning? The irrational number is an extension of the idea of fraction: a *complex* number is an extension of the idea of '*real*' number. If 0 is the mid-point of a long line, you would be prepared to place *every* fraction somewhere along that line and then give place to every *algebraical* number, i.e. to any number which is a root of an algebraical number equation. All these point-numbers are in one line: what about the complex of points in a plane containing the line? These are the numbers in the *complex field* of which solutions of $J - M$ are particular instances. Try to verify that $-(1 \pm i\sqrt{3})/2$ are roots of L as well as $x = 1$; but you may find this difficult. $\pm i$ are, of course, roots of M as well as ± 1.

What then of N and P? N has $x = 1$ and P has no solution? But the larger x is, the nearer we get to a solution for P. Can we say $x = \infty$ is a solution of P?

Again we are meeting a different type of number. If N_0 is the number of natural numbers, and N_{even} and N_{odd} the number of even and odd numbers, we have it seems, $N_0 = N_{\text{even}} + N_{\text{odd}}$. But if three boxes contain respectively all the natural, even and odd numbers, it is possible to take out one of N_0 each time one of N_{even} and N_{odd} is picked out. Thus $N_0 = N_{\text{even}}$ and $N_0 = N_{\text{odd}}$. An *in-finite* set is one in which there is a 1 — 1 correspondence between the members of the set and members of some sub-set of it.

An admirable analogy which brings home the distinction between the finite and the in-finite was made by Bertrand Russell. A goods locomotive, with a train of trucks, starts to move. With each clank a truck is jerked into motion. More and more of the train moves until, at the last clank, the whole is in motion. After a finite number of clanks all the train is in motion. If there were an infinite number of trucks, the clanking would continue for ever and the whole train would never be in motion.

The first sentence in this book was, 'Mathematics must begin with counting'. We have seen how the set of natural numbers needs to be extended to that of integers—positive and negative whole numbers—and from there to the set of rational numbers—all integers and fractions—and thence to algebraic numbers—all rationals and irrationals which can be roots of algebraic equations. Thus far we are concerned with numbers in a 'line', and there are still further numbers in the line before the set of 'real' numbers is complete. The number system can be enlarged yet again to complex numbers in the plane on each 'side' of the line and there are also in-finite numbers which cannot be positioned in any line. The concept of number is expanded as the need for the expansion arises. Yet, there at the base, are the natural numbers 1, 2, 3, . . .

Any definition of the natural numbers will be sufficient if a beginning number can be defined and, also, some succession defined, so that each number may lead to the next.

Remembering that the empty set is the set with *no* members—not the set with nothing in it; that it is the set with only self-contradictory and therefore non-existent elements; and remembering that the empty set is *unique*, we may define 0 as the property of every member of the empty set. We may say $\emptyset$ defines 0. The set which contains *only* the empty set defines 1. The set containing the already defined elements, defines 2, and so on.

$$\emptyset \rightarrow 0$$
$$\{\emptyset\} \rightarrow 1$$
$$\{\emptyset, \{\emptyset\}\} \rightarrow 2$$
$$\{\emptyset, \{\emptyset\}, \{\emptyset, \{\emptyset\}\}\} \rightarrow 3 \ldots \text{and so on} \ldots$$

It is important to realise that a number is not the same as a number of things. The number 2 is a property possessed by every pair of things. The property may be expressed as a class or set. Imagine a room in which you store boxes: imagine every box contains just two things: that room can define 2. The number 2 is the class of all classes containing two elements.

The answer to the question, 'What is a number?', was first given by Gottlob Frege in 1884. It says in effect, 'A number is that which is the number of a set'. Try convincing yourself this is, indeed, a justified and logically proper definition!

From the natural numbers we may define the negative integers and zero by the operation of subtraction. The rational numbers may be defined as the ratio of two whole numbers. The irrational numbers fill the 'gaps' in the line of numbers to complete the class of 'real' numbers. Complex numbers are related to points in a plane, specified by a pair of real numbers. Thus the number labelling is enlarged and explained; and, indeed, extended further still, beyond this.

25. Π and e

I WAS POOR at History but always remembered tags like Disraeli's 'Leap in the dark'. One of the 'catch-phrases' of Mathematics which people often remember is Euler's

$$e^{\pi i} = -1$$

where $e = 2\cdot71828 \ldots$ a *measure of maximum growth*, $\pi = 3\cdot14159 \ldots$ the *ratio of the circumference of a circle to its diameter* and $i^2 = -1$. I remember viewing this when I first met it, unknowing, with a fine feeling of wonderment. It may seem to you mysterious and wonderful—or mere nonsense! When I knew a little more mathematics and understood the equation, the mystery was gone— but its neatness remains. The equation says something like this, 'If you walk one mile to the east and then turn left and walk round a semicircle, you will find yourself one mile to the west of your

starting point! $-(+1) = -1$ can be likened to a movement from
1 to -1 in the *line* of numbers. $e^{\pi i} \cdot 1 = -1$ is like a movement
turning from 1 to -1 in the *plane* of complex numbers. But the
purpose here is not to explain this; it is to say something about the
numbers π and e.

Both numbers are *transcendental*, that is, they are irrational but
cannot be roots of algebraic equations. As decimals they are
never-ending but of precise value: there is no pattern of digits
as with fractions, but unlike $\frac{1}{10}(30 + \sqrt{2})$, for example, there is
no algebraic equation with them as roots. They may be roots of
trigonometrical or logarithmic equations but of no equation such as

$$x^2 - 6x + 8 \cdot 98 = 0$$

however complicated. A root of this equation was given above; it
is 3·14 to 2 decimal places and π is 3·14 to 2 decimal places, but π
or e cannot be a root of such an equation.

e. I have called e a measure of maximum growth. Let me justify
this. Money 'grows' under Interest. The great Reformation debate
on usury was on just this point: should the inanimate commodity
money be allowed to 'breed' and 'grow'. As the psalmist puts it,
'He that hath not given his money upon usury . . . shall never fall'.

Suppose, as an exercise, we find out what happens if £100 is
lent at 100 per cent compound interest for a year! If there is a single
payment of interest, then the amount A in £ is $100 + 100 = 200$.
If the interest is paid six-monthly, so the number n of payments is
2, then $A = 100 + 50 + 75 = 225$.
The calculation is $(100 + 100/2) + (100 + 100/2)\frac{1}{2}$

$$= 100(1 + \tfrac{1}{2}) + 100(1 + \tfrac{1}{2})\tfrac{1}{2}$$
$$= 100(1 + \tfrac{1}{2})(1 + \tfrac{1}{2})$$
$$100(1 + \tfrac{1}{2})^2$$

Extending this, when n is 3, A is $100(1 + \tfrac{1}{3})^3$ or about 237; when
n is 4, A is $100(1 + \tfrac{1}{4})^4$ or about 244, and so on.

We arrive at the formula

$$A = 100(1 + 1/n)^n$$

when the interest is paid n times during the year. Payment of
interest at the end of each month brings the amount up to £261;
every week brings it to nearly £270! We are beyond practicality,

but 100 payments would produce £270 10s. and 1000 produce £271 14s. 5000 brings the amount to £271 16s. and clearly a limit is being approached, £100 times 2·718 so far. The limit is the limit of

$$(1 + 1/n)^n$$

as n becomes infinitely large.

It can be shown that the limit is the limit of the sum

$$1 + \frac{1}{1} + \frac{1^2}{1 \cdot 2} + \frac{1^3}{1 \cdot 2 \cdot 3} + \frac{1^4}{1 \cdot 2 \cdot 3 \cdot 4} + \cdots$$

which is readily calculated to any desired number of decimal places. For

	1	$= 1\cdot 000\ 000$ etc.
	$1/1$	$= 1\cdot 000\ 000$
div. by 2,	$1/1 \cdot 2$	$= \quad \cdot 500\ 000$
div. by 3,	$1/1 \cdot 2 \cdot 3$	$= \quad \cdot 166\ 666$
div. by 4,	$1/1 \cdot 2 \cdot 3 \cdot 4 =$	$\cdot 041\ 666$ etc.

and the result follows by addition when the required number of noughts appear to the left. Check for yourself that e is 2· 718 281 830 to nine places by dividing as far as 12.

e behaves exponentially and is often referred to as the *exponential*. This means that whereas,

$$e = 1 + \frac{1}{1} + \frac{1^2}{1 \cdot 2} + \frac{1^3}{1 \cdot 2 \cdot 3} + \frac{1^4}{1 \cdot 2 \cdot 3 \cdot 4} + \cdots$$

we have also,

$$e^2 = 1 + \frac{2}{1} + \frac{2^2}{1 \cdot 2} + \frac{2^3}{1 \cdot 2 \cdot 3} + \frac{2^4}{1 \cdot 2 \cdot 3 \cdot 4} + \cdots$$

and

$$e^3 = 1 + \frac{3}{1} + \frac{3^2}{1 \cdot 2} + \frac{3^3}{1 \cdot 2 \cdot 3} + \frac{3^4}{1 \cdot 2 \cdot 3 \cdot 4} + \cdots$$

Test this to a few places of decimals. In general,

$$e^x = 1 + \frac{x}{1} + \frac{x^2}{1 \cdot 2} + \frac{x^3}{1 \cdot 2 \cdot 3} + \frac{x^4}{1 \cdot 2 \cdot 3 \cdot 4} + \cdots$$

a function of great importance in the Differential Calculus and other branches of Mathematics. The gradient or slope of its graph is, everywhere, the same as the value of the function itself.

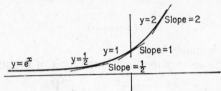

In the Integral Calculus, the size of an area of great importance, an area under the reciprocal curve $y = 1/x$, depends on e. The

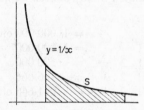

shaded area S, for instance, is given by $e^S = 4$, or S is about 1·386 sq. units.

The *Normal Distribution* depends on e. The shape of the curve $y = e^{-x^2}$ is the shape of natural happenings. Cannon balls fired to

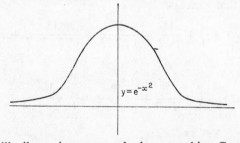

one spot will pile up in some such shape as this. Graphs showing the distribution of weights of babies at birth, or marks gained by candidates in a widely taken examination, will look something like this. The majority of weights or marks cluster about a central or *modal* value: fewer and fewer occur towards the extremes of high and low. A Mathematician was once defined as someone to whom it was obvious that the area under this curve was $\sqrt{\pi}$!

What is known as Whitworth's problem, states, 'If a quantity of fluid is shaken up and then comes to rest, prove that the chance that no particle of the fluid now occupies its original position is $1:e$'. You will probably find it easier to imagine a jar of sand instead of the fluid. Take two packs of shuffled cards and compare the playing cards one by one. What is the chance that no two are the same? The answer is similarly about $1:e$; a chance slightly greater than 1 in 3—much less than most people would expect!

One man with a rope round a bollard can exert a tremendous force. Why? It is because the friction increases *exponentially* as the angle the rope is turned through. Three turns about the post or bollard will *cube* the tension produced by one turn. The ratio may be as much as $e^{3\pi}$ or about 12,000:1!

π . π is the point in the number line reached by spreading out the circumference of a circle of unit diameter. For practical purposes of mensuration many approximations have been used. In the building of King Solomon's Temple the ratio is taken as 3, 'ten cubits from brim to brim, round in compass . . . and a line of thirty cubits did compass it round about'. Better values are $\sqrt{10}$, 256/81, 22/7, $3 + \frac{1}{8} + 1/60$ and 355/113. The last is correct to six decimal places. Archimedes was able to say that the ratio lay between 3 10/71 and 3 1/7. Values for π found by experiment are sufficient for practical purposes but add nothing to mathematical understanding.

Archimedes' method was to consider a circle as the limiting shape of a regular polygon. Try this yourself. If you have a little trigonometry you should be able to work out that the perimeter of an n-sided regular polygon in a unit-radius circle is $2n \sin 180/n$. The ultimate value of this, as the polygon gets nearer and nearer to a circle, is 2π. So $n \sin 180/n$ tends to the value of π as n gets larger. Try it for $n = 6, 10, 20, 90 \ldots$ Archimedes, in the third century BC used a geometrical form of the method. Formalised Trigonometry belongs to the seventeenth century and later.

The seventeenth century saw the beginnings of the calculation of π from the sum of a series of numbers. The simplest of these comes from Gregory's series obtained in 1671

$$\frac{\pi}{4} = 1 - \frac{1}{3} + \frac{1}{5} - \frac{1}{7} + \frac{1}{9} - \cdots$$

but a vast number of terms of this series are needed to reach a few correct decimal places in the value of π. After twelve terms only the first figure is correct! Modification of the series, however, allowed the calculation to 100 places of decimals by the early eighteenth century. Later Euler used,

$$\frac{\pi}{4} = \frac{1}{2} - \frac{1}{3} \cdot \frac{1}{2^3} + \frac{1}{5} \cdot \frac{1}{2^5} - \frac{1}{7} \cdot \frac{1}{2^7} + \cdots$$
$$+ \frac{1}{3} - \frac{1}{3} \cdot \frac{1}{3^3} + \frac{1}{5} \cdot \frac{1}{3^5} - \frac{1}{7} \cdot \frac{1}{3^7} + \cdots$$

With some careful tabulation of the working you can use this for your own calculation. Working to 4 dec. places and taking 5 terms of each series, you should get π to 4 figures.

Shanks, in 1873, using a quicker series, after prodigious labour, finally reached 707 places! As recently as 1945 there was still doubt about the accuracy beyond 500 places, but computer calculation produced 2,035 digits in 1947 and in 1955 no fewer than 3,089 places in 13 minutes of computer time. By 1957 the value had been found and checked to 10,000 places and in 1961 was calculated to 100,000 places in 9 hours.

The number π is so fundamental to mathematical analysis that to meet it first in circle measurement is perhaps unfortunate. This view of π—whose value is intriguing rather than important in exactitude—seems far removed from angles, probability, complex numbers, and such expressions as

$$\frac{2}{\pi} = \frac{\sqrt{2}}{2} \cdot \frac{\sqrt{2 + \sqrt{2}}}{2} \cdot \frac{\sqrt{2 + \sqrt{2 + \sqrt{2}}}}{2} \cdots$$

The connection with angle is easy, however. If you turn round completely, every separate part of your body moves through a distance which is 2π times the distance of that part from the axis of turning; 2π *is the measure of a complete turn.* π is the measure of half a turn: it occurs in Euler's equation in this sense. π, then, is expected where angles, trigonometry, complex numbers, are involved. You will find it much more surprising to learn that Insurance formulae for Life Expectancy involve π, or that if two whole numbers are chosen *at random* (there's the rub!) the chance that they are prime to each other is $6 : \pi^2$. It is reported that from an experiment with 50 students choosing numbers, a value of $3 \cdot 12$ was found for π, but this seems a lucky accident! The theory behind such probability assumes the selection of numbers continues *for ever.* The chance that a random chord of a circle cuts a given diameter is $2:\pi$. You might try to devise an experiment for this.

A matchstick is dropped upon the floorboards. What is the chance that the stick lies across a crack? This is an example of the naturalist Buffon's famous problem, solved in his 'Essai d'Arithmétique Morale', 1777. A plane is ruled with a system of equidistant parallel straight lines. A thin rod is thrown at random upon the plane. What is the chance that the rod will cut a line? The answer is, 'double the length: $\pi \times$ the width'. You can simplify the experiment by dropping a needle on a series of lines drawn so that their distance apart is the length of the needle. The chance is then $2:\pi$. Try it! If the needle is dropped, again many, many times, upon a square mesh or grating of parallel lines, their distance apart as before, the chance of a cut is $3:\pi$. See what you can get from this, also.

Finally, make a guess at the size of the following huge number, *factorial* 100, namely,

$$100 . 99 . 98 . 97 \ldots 5 . 4 . 3 . 2 . 1$$

A formula which gives a good approximation leads to

$$\left(\frac{100}{e}\right)^{100} \sqrt{2\pi \, 100}$$

This equals $(36\cdot79)^{100}(25\cdot07)$ or about $9\cdot4 \times 10^{157}$, that is 9 and 157 noughts! Write 10 instead of 100 in Stirling's expression above and compare the result with the multiplied value of factorial 10, namely $10 . 9 . 8 . 7 . 6 . 5 . 4 . 3 . 2 . 1$.

e and π are transcendental numbers for ever appearing in mathematical investigations. Lest you should think these the only transcendental numbers, let it be said that such numbers are more common than any others. Every number of the type a^b is transcendental if a is an algebraic number and b an irrational one. But you may continue to think π and e very, very special!

26. What the Calculus is about: (a) Differential

ZENO'S paradoxes are concerned with the nature of motion, an attempt towards bringing to logical clarity the investigation of change of position. In the fifth century BC he was provoking thought with his remarks about an arrow in flight. If an object is moving it is not in one place. But at any instant the arrow in flight is in one place: therefore, it is stationary. The argument is not silly: try answering it convincingly! The difficulty lies in the assessment of an in-finite number of positions taken in a finite time. In some respects the answers were not complete until Cantor's analysis of in-finite numbers just about a hundred years ago, but the ability to handle the infinitesimally small parts of *change* we owe to Newton and Leibnitz. The use of the Differential Calculus is relatively easy whereas a true understanding of it is far from easy.

Isaac Newton in England and Gottfried Leibnitz in Germany independently investigated what we call the Differential Calculus during the second half of the seventeenth century. Newton's publication of his 'Principia' in 1687 was largely due to Halley, well known for 'his' comet. The acrimonious controversy that arose between the two mathematicians was started and artificially fostered by others. There is a mass of biographical and historical material of interest about these two great figures, the German mathematician and logician, philosopher and diplomat, and the Englishman, perhaps the greatest mathematician of all time, certainly the equal of Archimedes and with only Gauss in some ways comparable with him, the Natural Philosopher, the M.P. and Master of the Mint, who esteemed his theological writings and Unitarian beliefs above his other work and who saw himself as he said, '. . . like a boy, playing on the seashore . . . while the great ocean of truth lay undiscovered before me.'

The nomenclature we now use is due to Leibnitz: Newton's own name for his investigation of infinitesimals was 'Fluxions'. He said he got the hint for his method from 'Fermat's way of drawing tangents'. Fluxion is a good and helpful name; Fluxions is a study of quantities in a state of flux, in a state of change.

A graph is a picture of change. (For convenience here the axes are labelled y and x, where y *depends upon* x, but any interpretation

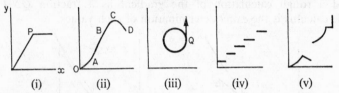

(i)   (ii)   (iii)   (iv)   (v)

may be given to y and x, such as distance, time, temperature, pressure, volume or speed.) In (i) there is first a steady change and then no change of y with respect to x, the slope or *gradient* is constant and then zero. (ii) might refer to a car journey with distance y plotted against time x. From 0 to A there is a gradual increase in the amount of change of distance, the speed is increasing, the gradient becoming steeper. From A to B the change is steady, the gradient constant. Now *you* describe the behaviour from B to C, at C, and from C to D. In (iii) no part of the changing pattern is steady; there is change of direction or gradient between any position and an adjacent one: but nothing 'unexpected' happens, the change is always 'smooth'. This is not so in (iv) or (v). The former might represent taxi fares; the meter jumps suddenly from one value to another; there is no smooth change. In (v) there are unexpected changes, gaps and jerks!

Suppose a little insect walks around curve (iii). At any instant he has a definite direction although it changes at every instant. At Q he faces as shown, he faces along the arrowed *tangent*. Between B and C in (ii) *every* speed from that at B to zero at C is gone through; the speed at any instant is the gradient of the tangent at that point. Fermat's hint concerned the finding of such tangents. (i) cannot be a distance-time graph: the sloping line shows a steady speed; the flat line, a state of rest; but what happens at P? No speed can change instantly; every intervening speed must be gone through, no matter how quickly. A car accelerating from rest to 60 mph. in 7 seconds (as an advertisement says) has an instant when it is going at walking pace!

(ii) and (iii) are graphs of *continuous* functions: the curves are smooth: any two points have a predictable mid-point on the curve:

at every point a tangent can be drawn and its gradient is the measure of the change at that point. That is what Fluxions is about, finding the measure of a smooth change. Careful placing of a ruler along a curve will give some idea of the direction of the curve at any point and a rough calculation of the gradient as a fraction QN/PN. The calculus is the *exact* determination of such values.

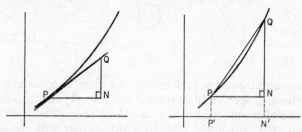

The change from P to Q *along the tangent* given by the fraction QN/PN is a measure of the change *at* P. Suppose Q is taken *on the curve* a little beyond P. As the curve is continuous we know Q can be brought nearer and nearer to P—there is always, for instance, the point mid-way along the curve, between P and a previous position of Q. The fraction QN/PN is now no longer the gradient at P but can be made nearer and nearer to it by gradually bringing Q closer and closer to P. Suppose Q is first chosen so that PN is one unit, then moved in so that PN is ·1, then ·01, then ·001 and so on. The calculated value of the fraction gets progressively nearer to the gradient of the tangent. If P is the point $(1, 1)$ on the curve $y = x^2$, we have $PP' = NN' = 1^2 = 1$, and:

PN	1	·1	·01	·001	·0001
$\dfrac{QN' - NN'}{QN}$	$\dfrac{2^2 - 1}{3}$	$\dfrac{(1·1)^2 - 1}{·21}$	$\dfrac{(1·01)^2 - 1}{·0201}$	$\dfrac{(1·001)^2 - 1}{·002001}$	
QN/PN	3	2·1	2·01	2·001	2·0001

The table can be continued for ever. For ever, PN can be reduced stage by stage to a tenth of its former length; for ever, a fraction results, closer and closer to 2. The gradient *is* 2, the *limit* of the fraction. It is never actually reached in the calculation, Q never actually gets to P, but by a process which can be repeated for ever Q comes as close to P as desired; the calculation comes as close as desired to 2.

PN is a small positive increase in x. Leibnitz denoted this by Δx. QN is the corresponding change, not necessarily small, in y, similarly labelled Δy. The fraction thus becomes $\Delta y/\Delta x$ and we write,

Measure of change of y with respect to x at P

$$= \text{gradient at } P$$

$$= \lim_{\Delta x \to 0} \frac{\Delta y}{\Delta x} = \frac{dy}{dx}$$

Δx and Δy are actual quantities, $\Delta y/\Delta x$ is an actual fraction, but $\dfrac{dy}{dx}$ is just *one* symbol for *one* quantity, namely the limit to which the fraction tends.

The concept of a calculable measure of smooth change is now established. You may well meet the symbols in reading which is not specifically mathematical. See if you can interpret verbally and understand the following statements and can say if they are justified, or give the conditions for which they are justified. The quantitive units of a *derivative* $\dfrac{dy}{dx}$ are those of the originating fraction. (The symbol k is to be read as a constant.)

1. $\dfrac{dy}{dx} = 0$. This means y has no change with respect to x.

2. $\dfrac{dp}{dq} = k$.

3. $\dfrac{dy}{dx} = y$.

4. $\dfrac{d(\text{area of circle})}{d(\text{radius})} = \dfrac{d}{dr}(\pi r^2) = 2\pi r$ sq. units/unit. This means that a measure of change of circle area with respect to its radius is numerically the value of the circumference. It makes sense if you consider a slight increase of area, implying the addition of a little 'thick' circumference.

5. For a sphere $\dfrac{dV}{dr} = 4\pi r^2$.

6. $\dfrac{d}{dt}$ (area of ink blot) = 2 sq. in. per sec.

7. If P is population, then at present $\dfrac{dP}{dt} = 190{,}000$ people per day.

8. Newton's Law of Cooling. θ is the temperature of the cooling body, t time, and a the constant surrounding temperature.
   $-\dfrac{d\theta}{dt} = k(\theta - a)$.

9. s is distance, v speed, f acceleration, t time.

   (i) $\dfrac{ds}{dt} = t - 5$.

   (ii) $\dfrac{dv}{dt} = f$ means the rate of change of speed is acceleration.

   (iii) Newton used dots to represent Fluxions. We use them with respect to time. Interpret $\ddot{s} = \dot{s} = s$.

10. $\dfrac{d}{dx}(uv) = u\dfrac{dv}{dx} + v\dfrac{du}{dx}$.

11. $\dfrac{dz}{dy} > 0$.

12. If A represents the warlike potential of country 'a' and B that of country 'b', interpret, $\dfrac{d}{dt}\left(\dfrac{dA}{dt}\right) \propto \dfrac{dB}{dt}$.

13. T is tension in yarn, z a vertical height and s a length of wool. $T\dfrac{dz}{ds} = k$.

14. V is the number of victims of an epidemic, and R the number of those recovered. $\dfrac{dR}{dt} = \dfrac{V}{4}$.

15. $\dfrac{dp}{dq} = 1/p$ means that provided p depends upon q as a continuous function, then the measure of change of p with respect to q is the reciprocal of the value of p.

The concept of a calculable measure of smooth change has been established but no general form of result has yet emerged. Returning to the graphs, let us suppose P is the point where $x = a$ on the curve $y = x^n$, n being a positive whole number. Let $x = b$ at nearby Q. We have,

$$\frac{dy}{dx} = \lim_{\Delta x \to 0} \frac{\Delta y}{\Delta x} = \lim_{b \to a} \frac{b^n - a^n}{b - a}$$

Now $\qquad (b^2 - a^2)/(b - a) = b + a$

and $\qquad (b^3 - a^3)/(b - a) = b^2 + ba + a^2$

and $\qquad (b^4 - a^4)/(b - a) = b^3 + b^2a + ba^2 + a^3 \ldots$

so we arrive at,

$$\frac{dy}{dx} = \lim_{b \to a} (b^{n-1} + b^{n-2}a + b^{n-3}a^2 + \ldots + ba^{n-2} + a^{n-1}).$$

If we allow Q to reach P, or b to equal a,

then $\qquad \dfrac{dy}{dx} = a^{n-1} + a^{n-1} + a^{n-1} + \ldots + a^{n-1} + a^{n-1}$

$$= n \cdot a^{n-1}$$

Whence the measure of change of x^n at $x = a$ is $n \cdot a^{n-1}$: a general result has appeared.

If a quantity y depend upon x . . . distance upon time, pressure upon temperature, cost upon labour, height upon speed . . . whatever it is, then y is a *function* of x: if the function is *continuous* then it is possible to *derive* another function which measures the change of the function. From the function $y = x^n$ we get the derived function or *derivative* $y_1 = n \cdot x^{n-1}$. y is just the name of the function, the y of the y axis of the graph. Nothing is lost in the understanding of Man if each man is labelled Smith! y_1 is derived from y as characteristics of the Smith family may be said to derive from Smith! The function derived from y_1 is y_2 and so on.

Newton used dots for differentiation and we use them when finding a derivative—the operation of *differentiation*—with respect to time. Change with respect to time is, of course, rate; rate of change is change with respect to time. Be careful you do not make the common error of confusing a measure of change with rate of change. If s refers to distance, v to speed and f to acceleration, convince yourself that in Newton's notation, $f = \dot{v} = \ddot{s}$.

Every smooth changing relation has a derivative that measures its change and that change has a derivative which measures *its* change, and so on. The function e^t depends on t, or we can speak of the *mapping* of t onto e^t: the relation is such that the measure of change equals the related value itself. An oscillation depending on an e^{-kt} factor will 'damp' or reduce: the exponential cannot be got rid of because it measures its own change!

A spherical toy balloon increases its radius from 10 cm. to 10·1 cm. in a warm room. By how much has the volume increased? The formula is $V = \frac{4}{3}\pi r^3$. Work it out! . . . But the measurement of change is $\frac{dV}{dr} = 4\pi r^2 = 400\pi$ cc/cm. *when* $r = 10$. Now $\Delta r = \cdot 1$, so

$$\frac{\Delta V}{\Delta r} \doteqdot \frac{dV}{dr} = 4\pi r^2 \Rightarrow \Delta V \doteqdot 400\pi \,.\, \Delta r = 40\pi = 126 \text{ cc. approx.}$$

Compare this easily obtained result with your other!

If you drive at 31 mph. your distance-time relation is $s = 31t$; your derived relation is $\dot s = 31$, your speed; $\ddot s$ your acceleration is zero, as is $\dddot s$ the rate of change of acceleration. Differentiation evaluates the measure of change in position, speed, acceleration, slope, electricity, waves, . . . and even life insurance!

Newton's Laws of Motion are still assuredly the fundamentals of Mechanics. Motion involves two quantities, amount of matter or *mass*, and speed in some direction or *velocity*. Motion is measured by their product mv, mass × velocity. If there is a change in motion, a *force* has produced the change. Newton's second Law states that the force producing the change depends upon $\frac{d}{dt}(mv)$ . . . so simple, so fundamental, so far reaching is this, that the solar system itself very, very nearly conforms to it!

A graph is a picture of change. Remember y and x can stand for anything but y must be related to x as a smooth curve for simple interpretation in the Differential Calculus. What are you told by the following information?

   (i) $y > 0$.
   (ii) $y_1 < 0$ but is zero at $x = 0$, $y = 1$.
   (iii) $y_2 > 0$ when $x < 0$ or $x > 10$, and $y_2 < 0$ when $0 < x < 10$.

x increases from L to R: positive gradient $y_1 > 0$ means y increases with x, the graph goes 'up'; negative gradient $y_1 < 0$ that it goes 'down': positive change in gradient means increasing gradient

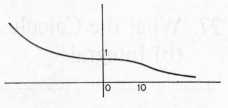

so that $y_2 > 0$ implies *turning* anticlockwise, and $y_2 < 0$ turning clockwise: $y_1 = 0$ means zero gradient and $y_2 = 0$ means zero turning. The information gives a very full picture of the shape of change though not of precise values.

A *local* maximum or minimum will always occur when there is zero gradient: the converse is not, however, necessarily true. Let us use this to solve a simple problem. A square-ended parcel is sent by post. How large can it be? The P.O. regulations require that the sum of the girth and length must not exceed 6 feet.

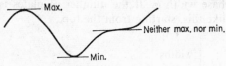

If the side of the square be w ft. and the length l ft. the volume V is given by w^2l with the proviso $4w + l \leqslant 6$.

$$\text{If } 4w + l = 6 \text{ then } l = 6 - 4w$$

$$\text{and } V = w^2 (6 - 4w) \text{ or } V = 6w^2 - 4w^3$$

For a maximum V the derivative must be zero,

$$\Rightarrow \frac{dV}{dw} = 12w - 12w^2 = 0 \Rightarrow w = 0 \text{ (min.) or } w = 1 \text{ (max.)}$$

With a width of 1 foot, the maximum dimensions of the parcel are 1 ft. × 1 ft. × 2 ft., a volume of 2 cu. ft.

The Differential Calculus deals with matters of motion, growth and change, rate and curvature, the shapes of loci and envelopes, and the determination of maximum, minimum and optimum values. Its notation is invaluable for statements concerning such matters, quite apart from mathematical theory or calculations depending upon it.

27. What the Calculus is about: (b) Integral

'INTEGRATION' means putting together to make a whole. A pile of square plates of equal thickness may produce a solid roughly a pyramid in shape; the more the plates and the thinner, the better the shape. Let us try to find a formula for the volume of a pyramid from this pile. Suppose the total height is h and base width b. If the number of plates is n, the calculation begins like this, starting from the top:

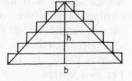

n	widths of plates as fraction of b	thickness as fraction of h	volume of plates as fraction of b^2h	total volume as fraction of b^2h
2	$\dfrac{1}{2}\dfrac{2}{2}$	$\dfrac{1}{2}$	$\left(\dfrac{1}{2}\right)^2\dfrac{1}{2}\ \left(\dfrac{2}{2}\right)^2\dfrac{1}{2}$	$\dfrac{1+4}{4}\cdot\dfrac{1}{2}$
3	$\dfrac{1}{3}\dfrac{2}{3}\dfrac{3}{3}$	$\dfrac{1}{3}$	$\left(\dfrac{1}{3}\right)^2\dfrac{1}{3}\ \left(\dfrac{2}{3}\right)^2\dfrac{1}{3}\ \left(\dfrac{3}{3}\right)^2\dfrac{1}{3}$	$\dfrac{1+4+9}{9}\cdot\dfrac{1}{3}$
4	$\dfrac{1}{4}\dfrac{2}{4}\dfrac{3}{4}\dfrac{4}{4}$	$\dfrac{1}{4}$	$\left(\dfrac{1}{4}\right)^2\dfrac{1}{4}\ \left(\dfrac{2}{4}\right)^2\dfrac{1}{4}\ \left(\dfrac{3}{4}\right)^2\dfrac{1}{4}$ $\left(\dfrac{4}{4}\right)^2\dfrac{1}{4}$	$\dfrac{1+4+9+16}{16}\cdot\dfrac{1}{4}$
5		$\dfrac{1}{5}$		$\dfrac{1+4+9+16+25}{25}$ $\cdot\dfrac{1}{5}$

114

We can now infer a formula for any value of n.

$$V = \frac{1}{n} \cdot \frac{1^2 + 2^2 + 3^2 + \ldots + n^2}{n^2} \cdot b^2 h$$

Now the numerator is S_2 of chapter 15, so

$$V = \frac{1n}{n} \frac{(n+1)(2n+1)}{6n^2} b^2 h = \left(\frac{1}{3} + \frac{1}{2n} + \frac{1}{6n^2}\right) b^2 h$$

A simpler table may thus be drawn up:

n	Volume as fraction of b^2h
10	$1/3 + \cdot 05 + \cdot 00166 \ldots$
50	$1/3 + \cdot 01 + \cdot 000066 \ldots$
100	$1/3 + \cdot 005 + \cdot 0000166 \ldots$
1000	$1/3 + \cdot 0005 + \cdot 000000166 \ldots$
20000	$1/3 + \cdot 00025 + \cdot 0000000004166 \ldots$
100000	$1/3 + \cdot 000005 + \cdot 0000000000166 \ldots$

The volume of the pyramid appears as a *limit*. As the number of plates becomes in-finitely large and their thickness in-finitely (infinitesimally) small, so the shape tends to a pyramid and the volume to the result $\frac{1}{3}b^2h$. To convince yourself of the argument, repeat the method for a cone made up of discs, and arrive at the formula $\frac{1}{3}\pi r^2 h$.

Archimedes in the third century BC was able to determine these volumes in some such manner as this. The result of the *integration* is a limit of a sum as it tends to an infinitely large number of infinitesimally small parts. Some questions immediately arise: Is the method justified? When can it be used? Is it always possible to find the required sum? Is it always possible to find the limit of the sum?

The answers are not easy. Archimedes had no answers although he obtained some brilliant results, notably that concerning equal areas on a sphere and on its circumscribing cylinder. Newton and Leibnitz established the processes and their consequences with magnificent insight and application, but Riemann's definition of an Integral belongs to 1854 and the *theory* of Integration only to the earlier years of the present century.

We were able to find the volume of the pyramid because it was possible to write down the size of successive plates. We need to know in what way successive *elements* of the sum behave. PQR is an arch. Can the area beneath it be found by integration? If the area be built up by successive elements such as those shown on the left, the resulting area will be too large; with those on the right, too small. With Leibnitz' notation, let the width of an element be Δx, the height of the arch y at the left of an element and $y + \Delta y$ at the right of it. Let ΔA be the area of the element of width Δx under the arch, then

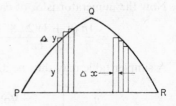

$$\text{rectangle } y . \Delta x < \Delta A < \text{rectangle } (y + \Delta y) . \Delta x$$

$$\Rightarrow y < \frac{\Delta A}{\Delta x} < y + \Delta y$$

The smaller the value of Δx, the more accurate the approximation to the area ΔA. In the *limit*, $\Delta y \to 0$ and $\dfrac{\Delta A}{\Delta x} \to \dfrac{dA}{dx}$

$$\Rightarrow y = \frac{dA}{dx} \dots \text{I}$$

The area A is the sum of all the elements A, or

$$A = \text{limit of Sum of all } \Delta A$$
$$= \lim S . (y . \Delta x), \text{ say,}$$
$$\Delta x \to o$$
$$= \int y . dx \text{ from } P \text{ to } R,$$

where the seventeenth century long S is used for the *limiting* sum.

If $x = p$ at P and q at Q and r at R, we have,

$$A = \int_{p}^{r} y . dx \dots \text{II}$$

meaning the area carved out by the march of the element from $x = p$ to $x = r$. y must depend upon x so that the pattern of successive elements is known, and for y to be $\dfrac{dA}{dx}$, A must be a *continuous* function of x. The curve PQ is expressible, in some way, in terms of x, as is QR: the *discontinuity* at Q does not spoil the validity of the integration; the elements may march from p to q and then from q to r,

$$\int_p^r y \, . \, dx = \int_p^q y \, . \, dx + \int_q^r y \, . \, dx$$

A curve does not have to be smooth to have an integrable area beneath it. All the graphs (i)–(v) of the last chapter have areas between them and the x axis. Some integrals can be evaluated at once as areas of triangles or rectangles. Look back to chapter 21 and then show that,

$$\int_{-1}^1 |x| \, . \, dx = 1 \quad \text{and} \quad \int_0^4 [x] \, . \, dx = 0 + 1 + 2 + 3 = 6$$

The sum tending to an integral may always be written down but it may not be possible to get a result from it; it may be impossible to find a limit to which it tends. Leibnitz regarded the Integral Calculus primarily in this sense, the finding of limiting summations, the determining of *Definite* Integrals: a Definite Integral is a *number*. Newton considered Integration primarily as the reverse of Differentiation: an *Indefinite* Integral is a *function*. Whether or not an Integral can be found, basically depends on this. An example will make this clearer.

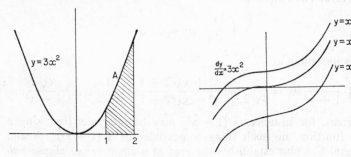

By I, an area such as A is given by,

$$\frac{dA}{dx} = 3x^2 \Rightarrow A = x^3, \text{ in this case, from } x = 1 \text{ to } x = 2,$$

$$= 2^3 - 1^3 = 7,$$

because x^3 *when differentiated* leads back to $3x^2$. We write, using II,

$$\text{the Definite Integral} \int_1^2 3x^2 \cdot dx = {}_1^2[x^3] = 8 - 1 = 7.$$

If $\dfrac{dA}{dx} = 3x^2$, A need not necessarily be x^3; it might be $x^3 + 1$ or $x^3 - 2$ or $x^3 + c$, any constant, for the measure of change of a constant is, of course, zero, and the derivative of all these expressions is $3x^2$. We write,

$$\text{the Indefinite Integral} \int 3x^2 \cdot dx = x^3 + c.$$

$x^3 + c$ is the function which satisfies the condition that the gradient $\dfrac{dy}{dx} = 3x^2$. The gradient of a curve does not depend upon the *position* of the curve with regard to the x axis, as the second diagram shows.

A Definite Integral may always be roughly determined from the required sum or graphical area, but the Indefinite Integral depends, absolutely, upon finding a function which, when differentiated, produces the function whose Integral is wanted. The subject is of great complexity, demanding knowledge and ingenuity: differentiation is a process of 'turning the handle' or 'pressing the button', whereas integration is a seeking into the dark, hoping to find results as simple as

$$\int \frac{1}{\sqrt{x}} \cdot dx = 2\sqrt{x}$$

or as intriguing as

$$\int \frac{1}{1 + x^4} \, dx = \frac{1}{4\sqrt{2}} \log \frac{x^2 + x\sqrt{2} + 1}{x^2 - x\sqrt{2} + 1} + \frac{1}{2\sqrt{2}} \tan^{-1} \frac{x\sqrt{2}}{1 - x^2}.$$

Whereas, for instance, $\sqrt{(1 - x^2)}$ may be integrated by using a sine function, no such thing is possible for $\sqrt{(1 - x^3)}$. Again, Integral Calculus establishes the area of a circle πr^2 or ellipse πab but shows there is no such formula for the circumference of an ellipse corresponding to $2\pi r$ for a circle; approximation, only, can be made to the sum. Integration gives $y = e^x$ as the solution of the *differential equation* $\dfrac{dy}{dx} = y$, but although another differential equation tells us the time of a wide swing of a pendulum, we can't find the time exactly . . . . can only approximate to it!

Differentiation delves into behaviour of change to determine its essentials. Integration tries to move outwards from change to determine its consequences. Distance and time changes contain within them the details of speed and acceleration. Change, under integration, moves outward to the consequences of change, acceleration to speed and distance, for example: thus, the differential equation $\ddot{s} = -ks$, soluble by integration, symbolically expresses the nature of all small oscillations, and

$$\frac{d^2u}{d\theta^2} + u = c \text{ (Polar coordinates, } u = 1/r, \ c \text{ constant)}$$

expresses the behaviour of bodies in space under Newtonian gravitation.

A Derivative expresses in mathematical terms the character of change: an Integral, as far as it is able, expresses the behaviour that follows change. The Differential Calculus digs into the roots of the growing plant: the Integral Calculus looks forward to the burgeoning.

28. Time and the Calendar

A GLANCE at the list of contents at the beginning of this book shows a host of obviously mathematical topics, headings unlikely, perhaps, to attract those without mathematical interest or ability, subjects like Arithmetic, Algebra and Geometry. The list could be very different and less obviously mathematical, but still concern parts of Mathematics. Surveying and map projection, Sound waves and music, Shape and aesthetic satisfaction, Biology and symmetry, Statistics and probability, the structure of Crystals, Traffic control, Navigation and astronomy, the Colouring of maps, Life Insurance tables, the theory of Games—all could appear but such applications usually require some basic thorough knowledge before interest can be used to advantage. The interest will then often develop to an understanding of the subject rather than to a deeper awareness of mathematical concept. Such topics are more for those already with some knowledge as well as interest. You may object that these are all too technical: Time and the Calendar concern us all; we all know something. Here is a topic all can pursue. What is Greenwich mean time? How is the length of a day affected by the Earth's leaning axis? Or by the position of the Earth relative to the sun? What and where is the International Date Line? Why and how does clock time change as one travels E to W or W to E? What is the history of Horology? The construction of a Sundial demands more than a little mathematics, accurate drawing and practical skill—a team effort, perhaps. Finding out about Sidereal time might be the concern of a physicist or astronomer. Where might the historian, for instance, come in? He could discourse upon the Roman Kalends, Nones and Ides, and the dating of the years A.U.C., discuss the Oriental year or the Byzantine year. He could write on the Mohammedan year of 354 days made up of 12 lunar months with the consequent movement of the fast of Ramadan through the seasons— eating and drinking is allowed only after sunset, which is very troublesome in Summer but much less so in Winter!—or the Jewish calendar with its intercalated month within the lunar system so that Passover remains in Spring. He could talk about the French Revolutionary calendar decreed in 1793; but he might prefer to say something about the Julian and Gregorian calendars, in this wise:

'The story begins when Julius Caesar, campaigning in Gaul, found that the expected issues of equipment to the troops did not suit the season of the year in which they were provided. The calendar was 'out of tune' with the Seasons: the 'year' was not measuring the cycle of the natural year: what could be reasonably said about the weather, temperature or climate at some part of one year, could not be said of the corresponding part of subsequent or earlier years. By BC 46 Julius had, as Emperor, the authority to deal with this. He took the advice of Sosigenes of Alexandria based on the astronomical observation of a solar year of $365\frac{1}{4}$ days. To bring the calendar back into line with the seasons he added 67 days to BC 46, 'the year of confusion', bringing it to 445 days in all!

The Roman year originally began on March 1st as our names

September, October, November, December testify. Julius Caesar devised 12 months, of alternately 31 and 30 days, from January 1st, but February with 29 days except every fourth year. A solar year of $365\frac{1}{4}$ days was thus achieved over every period of four years.

What was originally the fifth month was later named after Julius himself, as it still is, and Augustus Caesar took the next month to himself. It, however, gave him but 30 days of honour! To redress this, he added a day to August equating its length with that of July but upsetting the balance of the year, for the 'quarter' July, August, September would thus consist of 93 days. As a remedy, the month lengths were changed to what, so awkwardly, they are now; the 'quarters' amounting to 89 or 90, 91, 92 and 92 days.

The Ptolemaic year of $365\frac{1}{4}$ days is slightly too long, by 11 minutes and 14 seconds. This error multiplied year by year from the introduction of the Julian calendar. By AD 1582 Pope Gregory found it to be about 14 days, and he decreed Oct. 4th should be followed by Oct. 15th. It was the retrogression of Easter, based upon the Vernal equinox, which determined the matter, for Julius had established March 25th as the equinox and the beginning of Spring but this had shifted back to March 21st by the time of the Council of Nicaea in AD 325. It was at this council, called by the first Christian Emperor Constantine, that the dating of Easter was decided. Gregory returned the equinox to the 21st and not the 25th. His brilliantly simple solution to the Leap year problem—centuries 'leap' only when multiples of four, so 1900 was not a Leap year but 2000 will be—has an error which amounts to less than 1 day in 3000 years!

The Roman Catholic countries, of course, accepted the decree of the Pope in 1582, but the other Christian countries, Orthodox or Protestant, did not necessarily follow in revising their calendars. For a long time, our continent of Europe had both Old Style and New Style, as well as official and legal dating from monarchic accession, papal ordination and so forth. Scotland took the Gregorian calendar in 1600, but England not until 1752, with its cry, 'Give us back our 11 days!'. In the same year Parliament made January 1st the beginning of the year, in place of Lady Day March 25th, the Feast of the Annunciation just nine months before Christmas; March 25th the Vernal equinox of the Roman calendar and the beginning of Spring.

Japan did not change until 1873, nor China till 1912, but the countries of the Eastern Orthodox Church, such as Greece, waited until 1923. Mount Athos is still Julian!

So much for the year and months; the Mediaeval day was divided into twelve 'hours' of Light and twelve of Darkness. The Monastic sequence of worship was reckoned on a time scale from sunset to sunset—as the Church day still is, from Evensong to Evensong— and was far less arduous than it seems to us, reckoning by regular hours measured by a 'clock'. The diagram shows a complete cycle

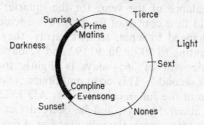

of day and night in 'summer' time when day 'hours' were long. In 'winter' time the day 'hours' would be short. The term 'o'clock' means what it says, reference to a clock—a church tower clock perhaps when they became fairly general. The hours then became standardized throughout the 24 and time measurement became established, not by Light and Darkness, but by the hours 'of the clock' in the cycle of two twelves in the 24.'

There you have a little sketch of our time and calendar. But what of the detail of Easter calculation as given in the Prayer Book? Of the Feriae and Dominical letters labelling the days of the week in relation to the date? There is much further material here for the historian. It was an early compiler of Easter tables, Dionysius Exiguus, who first dated the year from a supposed date of the Nativity of Our Lord, and called the year 247 from Diocletian, Anno Domini 532. The idea was taken up by Bede in the eighth century and subsequently spread to general use.

And what of the present? A calendar requires, ideally if rather clinically:

   (i) Days and dates that agree from year to year.
   (ii) Equal months beginning on the same day.
   (iii) Fixed days of the week for High Days and holidays.

Try composing your own!

29. Some Etymology, History and Biography

HWÆT! The great Old English poem Beowulf, partly historical but mostly fabulous, telling an exciting story of Danes and Sea Monsters in the 6th century, starts with this cry: a warning to be ready, to be still, to await the beginning of the recital of the tale. The word looks strange: what is its history? A poodle is but a puddle-hound. Why?

Words have two histories, the etymological and the semantic. Briefly, the etymological looks back to the origin of the word and the semantic to the sequence of change in meaning and usage of the word. The eighteenth century poet Alexander Pope, who re-marked, '. . . A little learning is a dang'rous thing . . . shallow draughts intoxicate the brain . . .', made himself most thorough in the English language, when a child, by reading through the Dictionary, with delight, at the beginning of the century. The obvious comment is, I suppose, 'How horribly precocious!', but the *reading* of a Dictionary can be great fun. Johnson's Dictionary of 1755, which cost him vast labour, 'dull work' he called it, is still fascinating to read, but a Dictionary such as the Shorter Oxford English Dictionary—OED—is now the place to look for word sources and usages.

The background of some Mathematical terms have been given already, as they arose, and a random dozen follow. Seek after more for yourself—there are many surprises awaiting your discovery!

ARITHMETIC	ME (Middle English)—OF (Old French)—Latin—Greek. Science of numbers. Confused in ME with Lat. ars metrica.
FARTHING	OE (Old English) Fourth part. cf. Yorkshire RIDING OE Third part.
FRACTION	ME—OF—Lat. *frangere* to break. cf. fragile.
ISOSCELES	Gk. *isos-skeles* equal legs.
KINETICS	Gk. to move. Study of motion. cf. cinema.

MATHEMATICS	Lat.—Gk. something learned.
MILE	Lat. *mille* thousand. 1000 complete paces, left-right-left, was originally about 1618 yards to the Roman soldier.
SCIENCE	ME—Lat. *scientia* from *scire* to know.
SCORE	OE—ON (Old Norse) *skor* a notch or cut: Icelandic twenty. To score is to cut. cf. Pork. To notch a stick cf. tally from F. *tailler* to cut.
SINE	Lat. *sinus* a bend, but used also for the bosom and mistranslated from Arab. *jaib* in this sense.
SURD	Lat. *surdus* deaf, from Gk. *a-logos* negative of *logos* number or word: used for irrational numbers not expressible as fractions.
VERTEX	Lat. highest or turning point: pointing to the apex of heaven: from *vertere* to turn.

Johnson compiled his Dictionary from his own vast knowledge and reading; it reflects his attitudes and prejudices, as the oft-quoted entry under *oats* testifies! The compilers of the OED from 1884 to 1928 used the services of thousands of readers and scholars who sent in postcards reporting the facts and dates of word-usages they had met with; in all, about 5 million quotations which resulted in 15,500 large quarto pages recording $\frac{1}{2}$ million words and some $1\frac{1}{2}$ million quotations. We can very often reach back to the past, to the source of the word before it came into English; the shape and form of the word may be enough: but the development of meaning we can only put together from research into use through the centuries. The history of a word consists of source and usage: much depends on *chance;* we can only piece together what we find to have happened.

Chance plays a fascinating part in all history—what if Cleopatra's nose had been long and sharp? To explore the history of Mathematics is to look for source and seek after development, to see the effect of the chance encounter with ideas and to find everywhere both the mutability and the continuity of the human genius.

The purpose, here, is to encourage *your* finding out something of the history of Mathematics. A number of approaches is possible: something may be found out about

    a period of Mathematics,
    the development of a particular topic,

the outline of the progress of a branch of Mathematics,
a chronology of great Mathematicians,
the life and work of an individual Mathematician,
the achievement in Mathematics of one nation or race,
or, some curiosities of Mathematics and Mathematicians.

There is enough for everyone: sufficient in the *lives* for those with little interest in the *work*. Creative Mathematics is predominantly the product of young men, but great Mathematicians have lived as varied lives as other men; few have lived dull secure lives; the delighter in anecdote will find a plethora of material.

In the early nineteenth century, the Norwegian Abel lived only to the age of 26 and the disillusioned young Galois was killed in a political duel in Paris before he was 21, but both had shown themselves Mathematicians of the first rank in their short unquiet lives. The power of Euler and Gauss extended into age: Euler was blind when he died at St. Petersburg at the age of 76 in 1783 but his memory and mental agility were still remarkable; Gauss was 77 when he died in 1855 after a lifetime of work; as a young man he had started to investigate circulating decimals by working out $1/n$ for the first thousand numbers! Both Blaise Pascal and Isaac Newton became deeply religious men, but in the previous century, the great Italian algebrist Girolamo Cardan was a dishonest though successful rogue. Archimedes lived a peaceful life until he got involved in the second Punic War in BC 212; Condorcet took an active part in the French Revolution, and during the Great War the French Mathematician Painlevé was for a time Prime Minister: such are the fortunes and abilities of some.

Scotland can boast among its Mathematicians, Gregory of the $\pi/4$ series, Napier and logarithms, Playfair and the parallel axiom, Simson the editor of Euclid's Elements, and Stirling of the $n!$ approximation formula. Ireland has William Hamilton (1805–65) who in his last years so preferred his work to his food that after his death the latter was sometimes to be found untouched among the chaos of his papers! Of the English, there are, for their several reasons, Barrow who relinquished his professorship at Cambridge to his pupil Newton, Halley, Wren, Wallis who also first studied the teaching of the deaf and dumb, Cayley and Sylvester, Charles Dodgson, George Boole, and in the present century they include Russell, Littlewood and Hardy, whose *apologia* should be read by everyone.

Seekers after anecdotes may find out about Tartaglia the stammerer, the dreams of Descartes, Fermat's marginalia or Hermite's hatred of exams. The staid seeker after chronology might try putting the following ABC of Mathematicians into date order! Apollonius, Bolyai, Ceva, Dedekind, Eudoxus, Fourier, Gödel, Hilbert, Isodorus, Jacobi, Klein, Lobachewski, de Morgan, Napier, Occam, Pappus, Quetelet, Riemann, Simpson, Thales, Urania, Vieta, Weierstrass, Ximenes, Young, Zeno. Add other claimants to the letters and raise objections to those listed, if you wish! The more overtly mathematical may trace briefly the early Mathematics of Babylon and Egypt, the establishment of logical method under the Greeks, the Mathematics of the middle and far East, its passage through the dark ages in Europe, to the middle ages and the spread of learning towards the Renaissance, and Mathematics' subsequent development measured by man, country or time, as the researcher prefers. There is also the search by topic. For instance, (i) the solving of equations: from Diophantus' solution of indeterminate problems, through the fun and games of Italian contests concerning cubic equations, to Abel's proof of the impossibility of solving algebraic equations above the fourth degree, by formula. (ii) The progress of trigonometry and logarithms running alongside the navigational needs of the seventeenth century. (iii) The history of mathematical symbols. (iv) The Mathematics of Harmony from Pythagoras onwards . . .

Wide is the opportunity whether your mathematical interest be great or small: use it as you will!

30. What is Mathematics?

THE 'Age of Reason' made use of the concept of the 'Noble Savage', a being untrammelled by the economical and political theories of the day and the dirt of consequent progress. We should not now consider man apart from his social environment, and few anthropologists would allow 'savage' tribes the attribute of primitive innocence. I suppose our equivalent of the eighteenth century ideal of a wise but innocent man would be a don manqué, a man of high IQ but no knowledge of the machinations of the world and society, developing his being, his body, mind and spirit in sufficient but lonely surroundings. His well-being would demand of him discovery about the earth and its fruits; his comfort would require basic competence in rude mechanics; his observation of the sea and sky would instruct him in natural forces. In the working of his intellect he could not, in his isolation, look to History or Language and Literature, nor probably concern himself with Art, although aware of Beauty: his mind could attempt to probe the mystery of Life and the ultimate of God, or exercise itself in the way of Mathematics. His study would be of Biology, Philosophy and Mathematics, and Mathematics, alone, can grow of itself.

This is, of course, as artificial as it is absurd! But the natural world around such a man would show him much that is a part of practical Mathematics, and the mental exercise of symbolic argument from chosen premises—if this be 'pure' mathematics—would free him from the verbal niceness so essential to philosophic argument. Certainly, a mathematician can never be without the opportunity of mental exercise; he can always make up a 'problem' and try to solve it.

What relevance has this to the nature of Mathematics? The 'savage', in so far as he had concern for Mathematics—not labelled as such—would only be interested in his own immediate practical need for it, as for instance the ancient Egyptians had, but the 'noble savage' would wish to develop a frame of argument within the mind, as the Greeks did. I am tying the early concepts of Mathematics to the convenient analogy: biologists find in the growth of animals the contracted history of a vastly long-span development. The hypothesised mental exercise which typifies Mathematics could be either of two types. Our innocently wise man could be *Logicist* and try to build a corpus of knowledge by logical movement upward and outward from one root fact such as a definition of number, or *Formalist* and state axioms of whatever nature the moment encouraged and then build logically upon them. Rousseau's ideal man is now become the mathematical *epistemologist* who has on his one hand the Academy of Plato and on his other the great mathematical philosophers of this century, Russell the logicist and Hilbert the formalist!

For Plato, Mathematics was ἐπιστήμη (epistēmē), Knowledge which leads through its ιδέα, Idea or ideal form, to an understanding of the True Being. In the Academy, Mathematics concerned the *relation* between *entities* but there was dispute what these were: were they 'real' as quantities and shapes are real? The query is still with us. Mathematics is concerned with the relations between entities which are themselves 'real' because they are mathematical. Mathematics deals with mathematical structures which, it then appears, sometimes reflect experienced reality. A recent writer says, 'Mathematics is simply what is produced by the research of mathematicians, who work within a tradition that has evolved and become consolidated during twenty-five centuries . . . Mathematics is what it is, and it needs no certificate of legitimacy from logicians or philosophers.' Russell defined Mathematics in 1903 as, '. . . the class of all propositions of the form "*p* implies *q*" . . . and *uses* a notion which is not a constituent of the propositions which it considers, namely the notion of truth'. Hilbert derived all that is Mathematics from the axioms at the root of each branch. 'Mathematics is what a mathematician does', becomes a useful, but unhelpful, compromise.

Russell and Whitehead's magnificent 'Principia', 1910–13, failed to reach far enough into the known world of mathematics, and Hilbert's approach proved too limiting, for in 1931 Gödel proved that our Arithmetic contained the seeds of its own inconsistency; that the consistency of Arithmetic cannot be established by the axioms on which it is based. He analysed the *undecidability* of problems arising from a set of axioms, that is, propositions which cannot be proved or disproved within the system. Neither the logicist nor the formalist could answer the final question as to the structure of Mathematics itself. Mathematics is greater than its own argument.

Gödel showed that such undecidable propositions within a consistently defined system may be demonstrated to be *true*, not by arguing within the system, but by arguing *about* it, by arguing *meta*-mathematically, as it is called. A deductive system is necessarily *incomplete* in itself: a computer cannot be capable of solving *all* problems!

In Information Theory, the analysis of message communication, there is a term *Redundancy* which measures the amount of a message which is not essential to its meaning. It would seem obvious that the best form of statement is one with no redundancy, but this is not so because in any human exchange of facts, whether by spoken word, directly person to person, or by telephone or wireless, by written word of hand or machine, by signs, signals or code, there is always the chance of error. The matter is not too bad if the presence of error is detectable, but it may well not be. 'Come at 2.00'. It is only after very careful inspection of spacing that one can see that the message should, perhaps, have read, 'Come at 12.00'. In some ways a poorer rendering would be better. 'Com eat z.00'. The recipient would certainly query this! A longer message with more redundancy, may be clearer in meaning even if retaining similar faults. 'Com eto see me at twelveo clock', is clear enough!

With the communication of ideas, rather than facts, errors of *interpretation* increase. An apparently clear exposition, simple and precise, without redundancy, may quite fail to convey what is intended. The title of this chapter is a question; a question is posed rather than an answer provided. Suppose, however, I write:

Question. What is Mathematics?

Answer. The analysis of structural pattern.

One person immediately visualises a mosaic tile arrangement, a second a picture of a damped oscillation, a third, the nine-times table, another thinks of the beauty of a sea shell, of a spiralling helix, or the petal symmetry of a flower; others will envisage at once the revolutions of an icosahedron or the possible shapes of crystals, will imagine bridge girders or air-flow diagrams, computer programmes or critical path problems; the mind of some might turn to atom behaviour, population distribution, cost control, projective geometry, electrical circuits, Boolean algebra or the design of wallpaper . . . No one mental response, or interpretation of the answer is what is intended: all—with *your* responses as well—all together might give some value to the suggestion. A composite answer containing a vast host of examples is an answer with a high degree of redundancy: yet *that* is the answer which might be enlightening.

There is, then, unlikely to be any conclusion to the discussion, 'What is Mathematics?', but the more you discuss and consider—no matter how inconclusively—the more you move towards some understanding of what an answer should contain. The industrialist will say one thing and the academic another; and not least will be the views of those who find in Mathematics just fun, beauty or delight.

G. H. Hardy in *A Mathematician's Apology* wrote, 'A Mathematician, like a painter or poet, is a maker of patterns. If his patterns are more permanent than theirs, it is because they are made with *ideas* . . . The mathematician's patterns, like the painter's or the poet's, must be *beautiful* . . . Beauty is the first test: there is no permanent place in the world for ugly mathematics.' Earlier, Weierstrass has said, 'A mathematician who is not also something of a poet will never be a complete mathematician.' But I feel sure the Industrialist will rarely agree!

Before the last war there was a great gulf fixed between the mathematician and the industrial *user* of mathematics. There is now a keen liaison between the mathematics of education and the mathematics of industry. I have heard Hardy dismissed as someone '. . . who stopped any mathematics he was doing as soon as he found it useful', and have heard mathematics listed as, 'Analysis of a real variable, Matrix algebra, Numerical methods, Probability and Statistics.' This does not debase Mathematics any more than the mechanics of perspective destroys—or creates—great painting. Mathematics is *determinist* not *stochastic*, it follows a reasoned pattern and does not depend upon a partially predictable chance; rather as the routing of cable to supply electricity to an estate is determinist but the timing of traffic lights at a busy junction to ease congestion is stochastic.

The industrialist needs mathematics as a *language of communication*, but probably does not need the mathematical analysis of the problem for an answer: the industrialist and the computer will find by experiment the answer which is practicable. In usages such as Queueing theory, Critical Path problems and Cluster analysis, this development may be sought . . .

But let that delightful, rambling, seventeenth antiquary Robert Burton of the 'Anatomy of Melancholy', have the last word, 'What more pleasing studies can there be than the mathematics . . .?'

Notes

Chapter 1

Some part of an answer to 'What is a number?' will be found in Ch. 24.

Details of the Rhind papyrus will be found in J. R. Newman (editor): *The World of Mathematics*: Allen and Unwin, vol. 1, part II. This compendium of mathematical writing is valuable to all who wish to read more.

Articles in encyclopaedias should be sought under the peoples or civilization wanted or under the heading 'Numerals'. Also see D. E. Smith: *Number Stories of Long Ago*: Ginn.

Chapter 2

I use the word *span* to refer to any complete set of counting digits.

The English word *cypher* derives from the Arabic *cifa*.

In Homer the verb *to five* means *to count*.

The Escorial library is in Spain NW of Madrid. The monastery and palace has an interesting history.

Arabic numeration came first into Europe through the Liber Abaci of Fibonacci in the 12th century. This and a mass of other information may be found in Boon: *Companion to School Mathematics:* Longmans.

It is a great help in understanding counting structure if the word *ten* is retained for the *span* symbol 10, and our own words used for the decimal notation and not for the Denary value. Those who dislike this, or find it confusing, may use digital wording—'My age is one-three-one years and two-zero months.'

I am using *decimal* to refer to any scale of notation, for any *value* of 10, but *Denary* when the scale is of our usual 10. Thus both Quinary and Denary counting is decimal, but decimal notation is not necessarily either of these.

Chapter 3

The distinction between decimal and Denary used here and above is not always made. I consider it a useful one.

It is clear from an example given that 12345 is a multiple of 5, in Scale Six. Why is this? Cf. our own divisibility test for 9. In what Scales will 12345 be divisible exactly by 5?

Chapter 4

Binary comes from Latin *bini* two by two.

Conversion from Denary to Binary is made shorter by turning first to *Octal* base 8 and then replacing each octal digit 0 1 2 3 4 5 6 7 by its binary equivalent 000 001 010 011 100 101 110 111. Thus *denary* 37 is *octal* 45 which is *binary* 100 101.

The example given for the human calculator needs seven human digits: choose other numbers as appropriate. 1101 is set on the 'machine' by activating the first, third and fourth 'digit'.

Here is an example of Denary subtraction by the *complement* method: $245 - 69 = 245 - 069 \rightarrow 245 + 930 = 1175 \rightarrow 176$ the right ans.

Chapter 5

Lewis Carroll's problems are being used here to introduce non-numerical algebraic symbols and not as an introduction to a formulation of Logic. See Ch. 9.

In Henry VIII Charter of 1546 the *Fellows* of Christ Church were subordinate to The Dean and Chapter of the Cathedral and were thus designated *Students*.

There is a reprint of *Symbolic Logic* published by Dover: New York and an interesting selection of problems in the Nonesuch Press *Lewis Carroll*, from which my examples are taken.

The Inverse is sometimes called the Opposite, and the Negative named Contrapositive.

It is accepted here that A'' or not (not-A), is the same as A.

The answers are:

   (i) Babies cannot manage crocodiles.
   (ii) No hedge-hog takes the *Times*.
   (iii) Guinea-pigs never really appreciate Beethoven.
   (iv) None but red-haired boys learn Greek in this school.
   (v) No kitten with green eyes will play with a gorilla.
   (vi) No cheque of yours, received by me, is payable to order.

Chapter 6

c, a, d_1, d_2 are all true for the 'New' Algebra, and c, a and d_2 for Number Algebra.

The Algebra chapters of W. W. Sawyer: *Prelude to Mathematics:* Penguin make admirable further reading to these chapters.

John Venn was a Cambridge logician 1834–1928. Some such diagrams were used by Euler from about 1770, but Venn's important extension of the idea dates from 1880.

A *class* is just another name for a *set*.

Here is a solution to the quadrilateral problem. Try to produce a better
one.

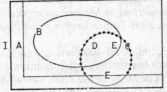

Chapter 7

$A \subset B$ means A is a sub-set of B, that A is *included* in B.

In Number Algebra $3 + 2 + 1$ is the same however worked out, but
$(3 - 2) - 1 \neq 3 - (2 - 1)$. Addition is both Commutative and
Associative but Subtraction is neither. What about Multiplication and
Division?

Answers: (i) A, (ii) A, (iii) $A \cup B \cup C$, (iv) $A \cap B$, (v) A, (vi) A, (vii) A,
(viii) $A \cap B \cap C$, (ix) A.

Possible labels are:

$\emptyset$

a	$(A \cup B)'$	ab	B'	abc	$A \cup B'$	$abcd$	I
b	$A \cap B'$	ac	$(A \cup B') \cap (A' \cup B)$	abd	$A' \cup B'$		
c	$A \cap B$	ad	A'	acd	$A' \cup B$		
d	$A' \cap B$	bc	A	bcd	$A \cup B$		
		bd	$(A \cup B) \cap (A' \cup B')$				
		cd	B	Many of these can be expressed in other ways.			

Chapter 8

64 read *Woman*.

The House Prefect problem is by K. Jalie. Answer: 21 *or* 24.

Coffee problem: x lb. of the 1st mixture and y lb. of the second lead to
$5x/6 + 2y/5 \leqslant 40$ and $x/6 + 3y/5 \leqslant 15$ with $3x + 2y$ to be a max.
Graphing suggests a solution (42, 12).

Chapter 9

{Sundays} $\subset$ {Rain days}
    Today ε {Rain days}
Today may or may not be a member of
Sundays. The syllogism is not valid.

{Pouches} $\subset$ {Marsupials}
{Kangaroos} $\subset$ {Pouches}
The syllogism is valid.

For traditional logic see encyclopaedia articles, Jevons: *Elementary Lessons in Logic*: Macmillan, or Ch. 1 of Kneebone: *Mathematical Logic and the Foundations of Mathematics:* van Nostrand. The quotation from Boole is taken from a long Note on him in Ch. 2, p. 51 of this work.

Look at a school 'Euclid' of 50–100 years ago and see for yourself how the free use of symbols has simplified the traditional school Geometry.

The Symbol for *or* is *inclusive* in the same way as $A \cup B$ means members of A or B or both. The symbol suggests itself by comparison with $\cup$ but has its source in Latin v for *vel*. An exclusive *or* is sometimes written v.

Question (ii) may be checked by considering, for instance, 'It is raining,' and, 'It is raining or snowing or both', *means* 'It is raining', at least. Alternatively it may be read as, for example, 'If p is true and also p or q or both are true, then p is true'. By analogy with Set Algebra, a Venn diagram will show that $P \cap (P \cup Q) = P$.

Number (viii) may be tested as, 'It is not raining and snowing', has the same meaning as, 'It is not raining or it is not snowing': alternatively as, 'If it is not true that p and q are the case, then p is false or q is false or both are false'.

Every relation involving the connectives $\vee \wedge$ has a *dual* in which these are interchanged, as are 0 and I when they occur. The dual of $(p \wedge q)' = p' \vee q'$ is $(p \vee q)' = p' \wedge q'$. These comprise de Morgan's Law. A proof of the first part is sufficient for both.

$$\text{By 1V } (p \wedge q) \vee (p \wedge q)' = 1$$
$$\text{but } (p \wedge q) \vee (p' \vee q') = (p' \vee q' \vee q) \wedge (p' \vee q' \vee p) \qquad \text{by II}$$
$$= (p' \vee 1) \wedge (q' \vee 1) = 1 \wedge 1 = 1 \quad \text{by IV}$$
$$\text{and also} \qquad (p \wedge q) \wedge (p \wedge q)' = 0 \qquad\qquad \text{by IV}$$
$$\text{but we show} \qquad (p \wedge q) \wedge (p' \vee q') = 0$$

whence it follows that

$$(p \wedge q)' = p' \vee q'.$$

To prove example (ix)

$$p \wedge (q \wedge r')' = p \wedge (q' \vee r) \qquad\qquad \text{by de Morgan}$$
$$= (p \wedge q') \vee (p \wedge r) \qquad \text{by Distr. law}$$

The relation $p \Rightarrow q$ is essentially an intuitive one. Mathematically $p \Rightarrow q$ is sometimes *defined* as $p' \vee q$: it properly has no meaning outside a defined one.

Anyone who wishes to go further with this Algebra will need to get used to symbols somewhat different from mine. The usage here emphasises the comparison with Set Algebra. A variety of symbols are used. Any

Boolean Algebra may employ the plus sign and juxtaposition for the two operators, so that, for instance, Distribution becomes $a(b + c) = ab + ac$ and $a + bc = (a + b)(a + c)$.

A very helpful little booklet to read is T. J. Fletcher: *0 and 1: Math. Pie*. For a full mathematical treatment see R. R. Stoll: *Sets, Logic and Axiomatic Theory:* Freeman. There are many examples in this book of the *use* of Sentence Logic in testing the consistency of an argument.

The relationship of logic and Boolean algebra to electrical circuits will be found in the booklet just mentioned, or in Goodstein: *Fundamental Concepts of Mathematics:* Pergamon, Ch. 4 and Andree: *Modern Abstract Algebra*: Holt, Rinehart & Winston, Ch. 3.

Chapter 10

$k(a) = (ka)$ is really an example of *addition*. For example

$$\begin{pmatrix} 1 & 2 \\ 3 & 4 \end{pmatrix} + \begin{pmatrix} 1 & 2 \\ 3 & 4 \end{pmatrix} = \begin{pmatrix} 2 & 4 \\ 6 & 8 \end{pmatrix}$$

If $|A| = 0$ then $\begin{pmatrix} a & b \\ c & d \end{pmatrix} \begin{pmatrix} d & -b \\ -c & a \end{pmatrix} = \begin{pmatrix} ad - bc & 0 \\ 0 & ad - bc \end{pmatrix} = 0 \cdot I$ or Z.

A is then said to be *singular*. You should notice it follows that $AA' = Z$ although neither A nor A' is Z. If $AB = Z$ it does not follow that $A = Z$ or $B = Z$ or $BA = Z$, but if B is non-singular then $A = Z$. For numbers, of course, $ab = 0$ requires a or b zero. With numbers, again, $ab = ac$ implies $b = c$ *unless* $a = 0$. In Matrix algebra $AB = AC$ implies $B = C$ if A is non-singular. The name for what I have labelled A' is the *adjoint* of A. The matrix $A^{-1} = A'/|A|$ is the *inverse* such that $AA^{-1} = I$. Cf. $xx^{-1} = 1$ for numbers.

To find a commonplace interpretation of matrix multiplication, look at the following bill:

4 lb. potatoes at 6d.	2	0
1 lb. butter at 3/6d.	3	6
2 tins baked beans at 1/-	2	0
2 pt. milk at 9d.	1	6
	9	0

and compare it with the product:

$$(4 \quad 1 \quad 2 \quad 2) \begin{pmatrix} \frac{1}{2} \\ 3\frac{1}{2} \\ 1 \\ \frac{3}{4} \end{pmatrix} = \begin{pmatrix} 2 \\ 3\frac{1}{2} \\ 2 \\ 1\frac{1}{2} \end{pmatrix}$$

For an easy introduction to this topic, see two booklets, Matthews: *Matrices*: Arnold.

Chapter 11

The equations $\begin{pmatrix} 3 & 2 & 8 \\ 1 & 1 & 3 \end{pmatrix} \begin{pmatrix} x \\ y \\ -1 \end{pmatrix} = 0$ may also be solved in matrix form

by the use of *detached coefficients*. You should compare this with the often-used method of *elimination* so as to understand it.

$\begin{pmatrix} 3 & 2 & 8 \\ 1 & 1 & 3 \end{pmatrix} \Rightarrow \begin{pmatrix} 3 & 2 & 8 \\ 2 & 2 & 6 \end{pmatrix} \underset{\text{by} \times 2}{\Rightarrow} \begin{pmatrix} 1 & 0 & 2 \\ 2 & 2 & 6 \end{pmatrix}$ by sub. $\Rightarrow$

$\begin{pmatrix} 2 & 0 & 4 \\ 2 & 2 & 6 \end{pmatrix}$ by $\times 2 \Rightarrow \begin{pmatrix} 1 & 0 & 2 \\ 0 & 2 & 2 \end{pmatrix} \Rightarrow \begin{pmatrix} 1 & 0 & 2 \\ 0 & 1 & 1 \end{pmatrix}$ and $\begin{pmatrix} 1 & 0 & 2 \\ 0 & 1 & 1 \end{pmatrix} \begin{pmatrix} x \\ y \\ -1 \end{pmatrix}$

$$= 0 \Rightarrow (x \quad y) = (2 \quad 1).$$

For many details of the use of matrices, see Andree: *Modern Abstract Algebra*, 5–11, and for *mappings*, 5–12; or refer to Harwood Clarke: *Modern Maths. at Ordinary Level*, Ch. 12: Heinemann. You will find it instructive to apply the transformations to the unit square made up by the points (0, 0) (1, 0) (1, 1) (0, 1). Matrix multiplication *maps* a point *onto* another: $x \rightarrow 2x$ and $y \rightarrow 2y$ in the first example; turns it anti-clockwise through the angle α about the origin, to its new position, in the last example. A mapping such as $x \rightarrow x + 2$, $y \rightarrow y - 1$ is done by the *addition* of matrices. $\begin{pmatrix} x \\ y \end{pmatrix} + \begin{pmatrix} 2 \\ -1 \end{pmatrix} = \begin{pmatrix} x + 2 \\ y - 1 \end{pmatrix}$

The final quotation is from OEEC: *New Thinking in Mathematics*, 1961, § 268.

Chapter 12

For the beginnings of Geometry see Coolidge: *A History of Geometrical Methods:* Oxford, pp. 1–23.

OED records no use of *geometer* before 1483. The reading of the Lombardic script is somewhat uncertain.

Comparison of the Pythagorean music scale and that of equal temperament, should be of interest to many.

The *Trivium*—from which we get *trivial*—consisted of grammar, rhetoric and logic.

The quotation is from Russell: *Hist. of Western Philosophy*, p. 233.

straight ME streȝt from OE streechen, to *stretch*.

strait ME streit from OF estreit, *tight*.

A shortest line on the Earth is a *geodesic*.

The terms *elliptic, parabolic, hyperbolic* are due to Pythagoras and are used in cases where a verbal distinction is needed, in some sense, between $\langle$ and $=$ and $\rangle$. Cf. the literary meaning of elliptic, parabolic and hyperbolic speech.

Where on an egg-surface are the angles of an equilateral triangle equal?
How do the angles vary as the triangle moves over the surface?

A résumé of Euclidean and non-Euclidean geometry can be found in
Kneebone: *Math. Logic*, pp. 134–7 and 182–3.

A facsimile of some pages of Descartes will be found in *The World of
Mathematics*, ed. Newman, vol. I. The date is given, misprinted as
1673.

You should find out for yourself something of the history and details of
the three Classical problems and of Zeno's four paradoxes.

Last paragraph of chapter: Cf. remarks of Oswald Veblen quoted in Bell:
Development of Mathematics: New York, p. 393. Veblen says else-
where, 'Geometry deals with the properties of figures in space'.

Geometry is now sometimes defined as the study of properties left un-
changed by certain transformations.

Chapter 13

For proofs of Euler's facts, see Rouse Ball: *Mathematical Recreations:*
Macmillan, Ch. 5 and 9.

Chapter 14

For more on Topology see, Barr: *Experiments in Topology:* Murray.
Lietzmann: *Visual Topology:* Chatto.

Chapter 15

1. $1 + 7 + 19 + 37 + 61 + 91 + 127 + 169 + 217 + 271 = 1000$.

2. Anyone able to perform simple integration may use the formula
$\dfrac{d}{dn} S_r = r \cdot S_{r-1} + B_r$, where B_r is a constant of Bernoulli, to find
expressions for other S. From S_3 he may find S_4 and so on.

$$\begin{aligned}
\text{Thus } S_4 &= 4 \int S_3 \, dn + B_4 n \\
&= \int n^2 (n+1)^2 \cdot dn + B_4 \cdot n \\
&= \int n^4 + 2n^3 + n^2 \cdot dn + B_4 n \\
&= \frac{n^5}{5} + \frac{n^4}{2} + \frac{n^3}{3} + B_4 n + C
\end{aligned}$$

Now $\qquad\qquad n = 0 \Rightarrow C = 0$

and $\qquad\qquad n = 1 \Rightarrow B_4 = -\frac{1}{30}$

Whence $\qquad\qquad S_4 = \dfrac{n}{30} (6n^4 + 15n^3 + 10n^2 - 1)$

3. $\begin{aligned}
S_{-1} &= 1 + \tfrac{1}{2} + (\tfrac{1}{3} + \tfrac{1}{4}) + (\tfrac{1}{5} + \tfrac{1}{6} + \tfrac{1}{7} + \tfrac{1}{8}) + \ldots \\
&> 1 + \tfrac{1}{2} + \tfrac{2}{4} \qquad\quad + \tfrac{4}{8} \qquad\qquad\qquad + \ldots \\
&= 1 + \tfrac{1}{2} + \tfrac{1}{2} \qquad\quad + \tfrac{1}{2} \qquad\qquad\qquad + \ldots
\end{aligned}$

$\qquad\qquad\qquad\qquad\qquad\qquad\qquad\qquad$ > any chosen number.

The sum of the reciprocals of the prime numbers is *divergent*, has no
limit, like the Harmonic series, but if all the terms containing the digit
9 are removed from the Harmonic series, the resulting series is *convergent*, it has a limit!

5. The *base* of natural logarithms is *e*. See Ch. 25.

6. The sum of all the reciprocals of the factors of a perfect number is
   always 2: thus for 6, $1 + 1/2 + 1/3 + 1/6 = 2$. More about such
   numbers can be found in Rouse Ball: *Mathematical Recreations*, Ch. 2.

7. There is extensive literature about tests of divisibility. For Duodenary
   digits T and E see the last part of Ch. 3.

8. The sum of odd numbers leads to a purely numerical demonstration
   of Pythagoras' theorem. For instance:

$$1 + 3 + 5 + 7 + 9 = (1 + 3 + 5) + (7 + 9) = 3^2 + 4^2$$
$$\text{and} \quad 1 + 3 + 5 + 7 + 9 = 5^2, \text{ by section one.}$$

For further reading see Beiler: *Recreations in the Theory of Numbers:*
Dover, and Reichmann: *The fascination of Numbers:* Methuen.

Chapter 16

By 'decimal' here and in the next chapter, is meant a decimal fraction, in
the Denary scale.

Mathematically the period p for a prime n is the smallest solution of the
congruence $10^p \equiv 1 \pmod{n}$. Try to find out the meaning of this.

Chapter 17

A denominator n obviously needs $(n - 1)$ forms of the decimal to provide
for all possible fractions. If the period is less than $(n - 1)$ there must be
other cycles to make up the total.

This topic is fairly easy at the present level but is difficult if probed. For
more, very mathematical, information see Barnard and Child: *Higher
Algebra:* Macmillan, Ch. 27, Hardy and Wright: *Theory of Numbers*:
OUP, Ch. IX or R. E. Green: *Primes and recurring decimals:* Mathematical Gazette No. 359. You will probably find it more fun to find
out about particular primes on a desk calculating machine.

Chapter 18

For much more of this subject see Rouse Ball: *Mathematical Recreations*,
Ch. VII, or Andrews: *Magic Squares and Cubes:* Dover. The Prime
square is quoted in Sierpinski: *Theory of Numbers:* Pergamon, p. 95.

Chapter 19

The *dexter* diagonal—like the Heraldic Bend—is from top left to bottom
right; the other is *sinister*.

To answer the query about the Distributive law, test

$$a(b + c) = ab + ac \quad \text{and} \quad a + bc = (a + b)(a + c).$$

'Ordinary' arithmetic of rationals is a Field: the arithmetic of integers, a Ring: the arithmetic of natural numbers, neither.

Chapter 20

$n!$ is *factorial n*, namely $n(n - 1)(n - 2)(n - 3) \ldots 3.2.1$. Thus 3! is 6 and 4! is 24.

The number of arrangements is 4 and 12 for orders 2 and 3. Rouse Ball quotes the results 56.5!.4! and 9408.6!.5! for orders 5 and 6.

Isomorphic is from Greek *equal-shaped*. Interchange of B and C converts III to I.

Felix Klein was a German mathematician 1849–1925.

In arithmetic mod 5 the multiplication table has the cyclic II pattern and in mod 4 the addition table is of type cyclic I. Another example of the Klein group consists of the sets $\emptyset$, A, B and $A + B$ combined under $+$, where this operation is defined as $A + B = (A - B) \cup (B - A)$.

A Latin square structure is not *necessarily* one that shows Group pattern. Consider the 5-square below. It represents a

```
A  B  C  D  E
B  A  E  C  D
C  D  A  E  B
D  E  B  A  C
E  C  D  B  A
```

non-associative structure: for instance $B(CE) \neq (BC)E$.

For more, see A. W. Bell: *Algebraical Structure:* Allen and Unwin, or Andree: *Modern Abstract Algebra,* Ch. 4.

Chapter 21

Some of this chapter is rather hard. If the drawing proves unpalatable, it may well be ignored. Two magnificent reference books for those who want more, are Cundy and Rollett: *Mathematical Models:* OUP and E. H. Lockwood: *A Book of Curves:* CUP.

Why are (x, y) co-ordinates called Cartesian?

Swastika is a Sanskrit word meaning Good Luck.

The centre of Apollonius' circle is midway between the points which divide $0_1 0_2$ internally and externally in the chosen ratio. $r_1 = r_2$ is simply the perpendicular bisector of $0_1 0_2$.

Cycloid (iv) is a diameter of the larger circle. If the moving circle rolls at a constant speed, the moving point performs Simple Harmonic Motion along the diameter!

Chapter 22

A *pencil* of lines is shown in the Polar co-ordinate section of the last chapter.

Drawings of envelopes, beautifully done, may be found in the *Book of Curves* already referred to.

The astroid is semi-cubical because the equation of the curve as a Cartesian locus contains the two-thirds power of x and y.

The parabola, ellipse and hyperbola are *Conic sections*. Find out the significance of this. They may be drawn as loci. If PN is the distance from a fixed line, and PS the distance from the fixed point S, then if $PS = ePN$ the locus of P is an ellipse if the *eccentricity e* is less than 1, a parabola if e is 1 and a hyperbola if e is greater than 1. See also the note to Ch. 12.

Chapter 23

Many 'proofs' of fallacies are common knowledge. I am aware of getting J from E. A. Maxwell: *Fallacies in Mathematics:* CUP, Ch. 10, and K is based on an example in Bradis, Minkovskii and Kharcheva: *Lapses in Mathematical Reasoning:* Pergamon. Both these books should be consulted by anyone wanting to go deeply into the sources of mathematical error. There are also examples of howlers of the type $\sqrt{(a^2+b^2)}$ $= a + b$ or, more comically, $\dfrac{26}{65} = \dfrac{2}{5}$.

C. Zeno of Elea c. 490–430 B.C. was the leader of the Stoic school of philosophy. His four famous paradoxes concern the nature of motion, the taking up of an *infinite* set of positions in a *finite* time. See E. T. Bell: *Men of Mathematics:* Penguin.

D. *AO* is the locus of points equidistant from AB and AC. *NO* is the locus of points equidistant from B and C. Congruency *AAS* may be used for 1 and 2, *SAS* for 3, and *RHS* for 4.

K is difficult. Are there the same number of points in every line? There is a little about finite and in-finite numbering in the next chapter. There is not a $1 - 1$ correspondence between points in the semi-circles and points in the diameters.

Chapter 24

Piet Mondrian, abstract painter, was born in Holland in 1872 and died in the USA in 1944.

If you have forgotten the quadratics formula, use:

$$\text{if } x^2 - 2bx + c = 0 \quad \text{then} \quad x = b \pm \sqrt{(b^2 - c)}.$$

An algebraic equation is one which is not, for instance, trigonometrical or logarithmic. Thus $2 \sin x = 1$ and $2^x = 3$ are *not* algebraic equations.

For a simple study of Cantor's trans-finite numbers, see: (American) *Mathematics Student Journal*, vol. 10, no. 4, or *The World of Mathematics*, pt. X, or Pedoe: *Gentle Art of Maths.*, Ch. III.

For Frege, see Kneebone: *Mathematical Logic and the Foundations of Mathematics*, Ch. 6 § 4. It is not easy.

For a full and readable treatment of *number* see, Bertrand Russell: *Introduction to Mathematical Philosophy:* Allen & Unwin.

Chapter 25

For the Elizabethan view of usury see L. C. Knights: *Drama and Society in the Age of Jonson*, pp. 164–8. The quotation is from the Prayer Book version of Psalm 15.

A quantity increases *arithmetically* by addition, *geometrically* by multiplication, and *exponentially* by raising to a power. Thus, e.g. 2, 4, 6, 8 . . . 2, 4, 8, 16 . . . 2, 4, 16, 256. . . .

$y = e^x$ leads to $x = \log y$ where the logs are to base e. These are the *natural* logarithms. You will find a table of them in most sets of 4 figure tables. $e^S = 4 \Rightarrow S = \log 4$.

The solution of Whitworth's problem is not difficult. It can be found in Edwards: *Integral Calculus*, vol. II, p. 796—there is no calculus in the working!

The Solomon's building references are I Kings 7.23 and II Chron. 4.2 and concern the making of a 'molten sea'.

James Gregory 1638–1675 was a Scottish professor of Mathematics. The series follows from the differentiation of the Inverse Tangent function.

For examples of the tabulated calculation of π see Boon: *Companion to School Mathematics:* Longmans, p. 56 et seq.

For π to 10,000 places and some details of calculation see, Oxford Math. Conference proceedings 1957: Times, p. 12 and footnote.

The infinite product for $2/\pi$ was obtained by the Frenchman Vieta 1540–1603 by geometrical means.

The $6:\pi^2$ experiment is mentioned by Rouse Ball.

Chapter 26

Latin *calculus* is a pebble used in calculation.

For Newton see *The World of Mathematics*, vol. 1.

If y is a differentiable function $f(x)$, the derivative $\dfrac{dy}{dx}$ is defined as

$$\lim_{h \to 0} \frac{f(x + h) - f(x)}{h}.$$

To be strictly accurate a differentiable function and a continuous function are not necessarily the same thing.

Mapping notation: $x \to x - 1$, for instance, represents $y = x - 1$. For exercises in Differential calculus see textbooks of Sawyer: *Mathematician's Delight:* Penguin.

Chapter 27

The method for finding a pyramidal volume applies equally whether the
solid is right or oblique.

An exercise which will test your perception and
ingenuity is to find the volume common to two
equal pipes intersecting at right angles. *AB* and
CD are two pipes of the same radius *r*. The
diagram shows the central section. The common
volume has a square section, diminishing in size on
either side of this central section. Try finding the
volume by summing thin square plates. The
answer is $\frac{8}{3} r^3$ . . . it does *not* contain π.

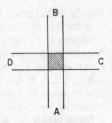

Find out for yourself the fact about equal areas on a sphere and its surrounding cylinder. The fact is beautifully simple but awkward to
describe.

Chapter 28

Sundials originally showed roughly equal divisions from sun-rise to sunset.

There is a short interesting article in Mathematical Pie no. 51. See also
A. P. Herbert: *Sundials Old and New:* Methuen. Also H.M.S.O.:
Time Measurement, two fine cheap booklets, I and II.

There is a useful pamphlet on Medieval Reckonings of Time: R. L.
Poole: S.P.C.K. A New Calendar is suggested in the Mathematical
Gazette no. 352, W. F. Bushell: *Calendar Reform*.

Chapter 29

The development of the English language is in three parts:

Old English—*not* Anglo-Saxon—roughly to 1100.
Middle English—roughly 1100 to 1400.
Modern English—roughly since 1400.

Hwæt has little history. It is an expletive, a call for silence by the *scop*
or reciting poet.
Poodle and *puddle* both derive from the same Dutch source.
The Pope quotation is from *An Essay on Criticism* 1709, pt. II.
The shorter OED is the place to look for further etymologies, but Boon:
Companion to School Mathematics contains a wealth of fascinating
information on etymology, history and biography.
Oats: 'A grain, which in England is generally given to horses, but in
Scotland supports the people.'

For racy and readable biography see E. T. Bell: *Men of Mathematics:* Penguin.

G. H. Hardy: *A Mathematician's Apology:* CUP, written early in the last war.

Occam and Urania are perhaps unfairly listed in the ABC! But Bertrand Russell refers to Occam's 'razor' in his own progress of logical development, and Urania, the Muse of Astronomy, must surely have Mathematics in her pagan charge!

Some histories of the subject are:

> Rouse Ball: *A Short Account of the History of Mathematics:* Dover.
> E. T. Bell: *The Development of Mathematics:* McGraw-Hill.
> T. L. Heath: *A Manual of Greek Mathematics:* OUP.
> F. Lasserre: *The Birth of Mathematics in the Age of Plato:* Hutchinson.
> D. J. Struik: *A Concise History of Mathsematics:* Bell

and of course, *The World of Mathematics.*

Chapter 30

Jean-Jacques Rousseau 1712–1778; great French writer and social philosopher.

Plato and the Academy: see Lasserre: *The Birth of Mathematics in the Age of Plato.* Raphael's well known fresco at the Vatican, executed 1510–11, *The School of Athens*, gives a not unhelpful view of the Academy as it might have been (Ib. pp. 36–7). The picture is reproduced in colour in *The Masters*, no. 41: Knowledge publications.

A synopsis of Russell's conception of Mathematics will be found in Kneebone: *Mathematical Logic*, Ch. 6: the quotations given are from this fine book.

Gödel: see *The World of Mathematics*, vol. III or *Math. Gazette*, no. 375.

There is an erudite note on crystalline structure and mathematics in Kneebone: p. 355.

The quotation from Weierstrass—with a crowd of others—is from the beginning of Bell: *Men of Mathematics.*

Francis Bacon in his comprehensive classification of Human Knowledge in *The Advancement of Learning* 1605 labels Mathematics *An Appendage to Speculative and Practical Philosophy* and splits it into *Pure:* Geometry, Arithmetic, Algebra, and *Mixed:* Perspective, Music, Astronomy, Cosmography, Architecture, Enginery.

Some books for further reading are:

R. L. Goodstein: *Fundamental Concepts of Mathematics:* Pergamon.
S. Körner: *The Philosophy of Mathematics:* Hutchinson.
F. Land: *The Language of Mathematics:* Murray.
W. D. Lewis: *Mathematics Makes Sense:* Heinemann.
D. Pedoe: *The Gentle Art of Mathematics:* EUP.
O. G. Sutton: *Mathematics in Action:* Bell.
D'A. W. Thompson: *On Growth and Form:* CUP (abridged).

Glossary

Abacus. A counting frame.

Associative law holds for the operation * if $A*(B*C) = (A*B)*C$, i.e. brackets are not necessary for the repeated operation. Cf. $+$ and $-$.

Axiom. A self-evident proposition presented without proof.

Base. Number *base* is the value of the *ten* in a decimal system. The *base* of a logarithm is the number whose powers provide the logs, e.g. $10^2 = 100 \Rightarrow \log_{10} 100 = 2$ and $2^3 = 8 \Rightarrow \log_2 8 = 3$.

Binary scale of notation is counting with 0 and 1. *Binary* algebra has symbols which can take only two values, such as the 0, 1 of Sentence logic.

Class. Set.

Coefficient, e.g. coefficients of $2x - 3y - 4z$ are 2, -3, -4.

Commutative law holds for the operation * if $A*B = B*A$. Cf. $+$ and $-$.

Complement of digit; sum of whole number and complement is the highest valued digit of the positional counting scheme used.

Complement of angle: sum of angle and complement is a right angle.

Complex numbers can be designated by a point in a *plane*. Cf. *real* numbers.

Congruence of numbers. $a \equiv b \pmod{n}$ means that a and b both have the same remainder when divided by n. Geometrical figures are *congruent* if they are exactly superimposable.

Coordinates. Numbers, number-pairs, triples . . . which specify the position of a point in a line, plane, 3D space . . .

Decimal. Counting in *tens*.

Denary is the usual counting system, with a two-hand *span*.

Derivative. The function derived by differentiation.

Digit. A single whole number symbol. (Latin *digitus* a finger or toe).

Disjoint sets have no common members, apart from $\emptyset$.

Distributive law holds for operation * *over* operation $\circ$ if $A*(B \circ C) = (A*B) \circ (A*C)$. Cf. $a(b - c) = ab - ac$.

Dual. A valid statement obtained from another by the interchange of certain symbols, words or phases.

Empty set ∅. The set with no elements.

Entasis. The swell of the shaft of an architectural column.

Epistemology. The theory of the grounds of knowledge.

Equivalent. ⇔ meaning '. . . implies, and is implied by . . .' *or* '. . . if, and only if, . . .'

Exponential. The value of *e* is 2·718 . . . See chapter 25 and Note.

Factor. An exact divisor.

Factorial. For integer n, $n! = n(n - 1)(n - 2)(n - 3)$. . . $3 . 2 . 1$.

Hieroglyph. A picture or figure standing for a word, syllable or sound.

Hypothesis. A proposition used as the basis of argument.

Idea. In Platonic sense: the archetype of which all individual things are imperfect copies.

Identity. An algebraic equality which is true for *every* possible value of the symbols.

Implies. ⇒ means, 'If . . . then . . .'

In-finite: as opposed to *finite*, e.g. not countable.

Integer. Whole number.

Inverse. Opposite, e.g. the inverse of + is −; of $A \Rightarrow B$ is $A' \Rightarrow B'$; of log. is anti-log.

Irrational numbers are real numbers which are not rational.

Isomorphism. Mathematical systems which possess the same *shape*.

Limit. A finite end to which a mathematical process may tend by repeated application, although it is never actually reached.

Mapping. Relating a point's position to another according to a stated rule.

Mensuration. The calculation of areas and volumes.

Multiple. If x is a factor of the number y, then y is a multiple of x.

Natural numbers. The counting numbers 1, 2, 3, . . .

Octal. Counting to base 8.

Postulate. A statement which is *assumed* to be valid.

Premiss or *premise.* A proposition from which others may follow.

Prime. A whole number which is not the product of others.

Prism. A solid with uniform cross-section throughout its length.

Proposition. The meaning of a sentence.

Quinary. Counting in fives.

Radix. The 'root' or base value in decimal counting; the value of *ten*.

Rational numbers are the ratio of two integers, the second of which is not zero: the integers and fractions.

Real numbers can be specified by a point in a *line*. Cf. complex numbers.

Scalar. A quantity which can be represented by a pure number.

Scale of notation. The specification of a decimal counting system, e.g. Quinary or Denary.

Sentence. A simple indicative statement which is entirely true or entirely false, or a combination of such statements, or their negatives, connected by *and* or *or*.

Span. The word used in this book for the number of symbols in a positional form of counting.

Surd. A number involving the signs $\sqrt{}$, $\sqrt[3]{}$ etc.

Ten. The word for the number 10 whatever its value.

Ternary. Counting in threes.

Unicursal: capable of being traced in one continuous movement.

Vector. A quantity requiring both magnitude and direction to specify it; needing a pair of numbers to determine it.